Sarah Spiekermann
Value-Based Engineering

Also of interest

Empathic Entrepreneurial Engineering.
The Missing Ingredient
David Fernandez Rivas, 2022
ISBN 978-3-11-074662-4, e-ISBN (PDF) 978-3-11-074682-2

Technology Development.
Lessons from Industrial Chemistry and Process Science
Ron Stites, 2022
ISBN 978-3-11-045171-9, e-ISBN (PDF) 978-3-11-045163-4

Robotic Process Automation.
Management, Technology, Applications
Edited by Christian Czarnecki, Peter Fettke, 2021
ISBN 978-3-11-067668-6, e-ISBN (PDF) 978-3-11-067669-3

Practical AI.
For Business Leaders, Product Managers, and Entrepreneurs
Alfred Essa, Shirin Mojarad, 2022
ISBN 978-1-5015-1464-7, e-ISBN (PDF) 978-1-5015-0573-7

Engineering Innovation.
From Idea to Market Through Concepts and Case Studies
Benjamin M. Legum, Amber R. Stiles, Jennifer L. Vondran, 2019
ISBN 978-3-11-052101-6, e-ISBN (PDF) 978-3-11-052190-0

Sarah Spiekermann
Value-Based Engineering

—

A Guide to Building Ethical Technology for Humanity

DE GRUYTER

Author
Prof. Dr. Sarah Spiekermann
Vienna University of Economics and Business (WU)
Welthandelsplatz 1
1020 Wien
Austria

Illustrations and figures by Marie-Therese Sekwenz, co-developed with Sarah Spiekermann; please acknowledge the free use of this book's illustrations as "VBE Graphics CC BY-NC-ND"
VBE Graphics CC BY-NC-ND

ISBN 978-3-11-079336-9
e-ISBN (PDF) 978-3-11-079338-3
e-ISBN (EPUB) 978-3-11-079350-5

Library of Congress Control Number: 2022947585

Bibliographic information published by the Deutsche Nationalbibliothek
The Deutsche Nationalbibliothek lists this publication in the Deutsche Nationalbibliografie; detailed bibliographic data are available on the Internet at http://dnb.dnb.de.

© 2023 Walter de Gruyter GmbH, Berlin/Boston
Cover image: Marie-Therese Sekwenz
Typesetting: Integra Software Services Pvt. Ltd.
Printing and binding: CPI books GmbH, Leck

www.degruyter.com

Preface

The journey of this book started nearly 10 years ago when I prepared my first textbook for students on Ethical IT Innovation in 2014. At the time few people realized how urgently we need to develop a more thoughtful and ethically sensitive approach to IT system design. Of course, many critical minds at universities and tech-focused NGOs had realized that IT systems would need to be more sensitive to privacy and security. The EU's General Data Protection Regulation (GDPR) was under way. But beyond the GDPR effort hardly any initiatives were able to successfully call for a more thorough protection of human values. This has changed dramatically since 2014. Since Edward Snowden's revelations on the scale of global digital surveillance (late 2013), Cambridge Analytica's exposure for manipulating the Trump election (2018) and the influential documentary film *The Social Dilemma* (2020), many noteworthy activists, whistleblowers, individual tech pioneers, corporate leaders and politicians have been critical of the sheer scale at which our IT innovations threaten democracy, undermine political and economic control, change the nature of warfare, and influence human lifestyle, personality and health. A political debate has emerged around untransparent, unaccountable and uncontrollable AI systems, which might compromise the values that we as a society hold dear. Furthermore, the looming specter of climate change has caused financial investors, insurance companies and other economic forces to recognize the need for a more environmentally friendly and social governance of IT operations.

But how are IT companies to build systems that live up to this new call for more value respect? In 2014, I realized that it is not easy to build IT systems in a value-based manner. At the time I had already been teaching IT system design and analysis at the University for 10 years. I had also been teaching about human-computer interaction, user-centric system design and user experience. But neither standardized system development lifecycle methods, nor more agile approaches to system design, nor HCI or UX coursework seemed to integrate any sensible methodological approach to address or even think about value issues. Standard textbooks on software engineering mostly ignore the social challenges of IT systems to this day. The only straw I could clutch at back then was the work that Batya Friedman as well as others had been doing on Value Sensitive Design (VSD) since the late 1990s. The case studies accumulated by the VSD community over the years as well as her conceptual work on values in system design inspired my own work on Value-Based Engineering as it is summarized in this book. In fact, I tried in my 2014 textbook on Ethical IT Innovation to come up with an "Ethical System Development Life Cycle," in which I combined the Value Sensitive Design approach with participatory design and risk management. But at this first attempt, I was not able to solve the core conundrum of value-based design; namely, how to go systematically from value principles to practice. Or in other words: how to practically, systematically and traceably derive concrete IT system features and requirements from lofty value principles such as love, dignity, freedom or fairness. It took the intervening 10 years for me to find a reliable answer for this problem, which you now find published in this volume.

How did I find the answer and develop Value-Based Engineering since 2014? I certainly did not do it alone. Together with a workgroup of IEEE volunteers, I engaged in a five-year-long standardization project called "P7000." I co-initiated, served as vice chair, and intellectually guided this project, which aimed to develop the world's first model process for tackling ethical concerns in system design, between 2016 and 2021. Given the enormous amount of ambition at the heart of this goal, it should not come as a surprise that endless and fierce discussions emerged. I can now with certainty attest to the fact that discussions around ethics and values and how to best live up them are among the most emotionally charged. They touch upon very personal human convictions and philosophies on how to see the world and do things. Workgroup members engaged, for example, in discussions about what the word "design" actually means, whether ethical systems should have "zero fault tolerance," what "values" are and how they can end up in a system since they are actually invisible phenomena. They debated how companies' pursuit of profit could at all be aligned with ethics, how top management could be involved more in a value mission, how work conditions for engineers could allow for more room to tackle ethical issues, how to bridge the gap between engineers and managers ("the two-domain problem"), how to respect human rights in a global standard if the whole world does not equally share in them, how to avoid ethical relativism, etc. I think that leading and completing the IEEE Std 7000^{TM} project has been one of the most painful and difficult endeavors of my life. As in any truly democratic and bottom-up decision process, the challenge is how to ensure the highest quality standards while listening to and including everyone. Is this at all possible? IEEE Std 7000^{TM} has been adopted by ISO/IEC JTC 1 and is also available as ISO/IEC/IEEE 24748-7000, *Systems and Software Engineering – Life Cycle Management – Part 7000: Standard Model Process for Addressing Ethical Concerns during System Design*. Personally, I am proud of the result. I believe that companies are well advised to use IEEE Std 7000^{TM} – or parts of it. Their IT systems will be much better than what they are currently building.

Why will IT systems be much better with the Value-Based Engineering approach? I think an eye-opener for me was the case studies we conducted at my university, while developing the standard. In particular, we explored how the use of the three grand philosophical streams of utilitarianism, virtue ethics and duty ethics would sensitize IT innovation projects for the likely value harms inherent in their projects as well as for the potentially positive value propositions dormant in the technology. What struck me most was how one and the same technology can actually be put to bad as well as good use depending on how it is built. We can build an IT system for food delivery that treats bike couriers like isolated, surveilled and controlled pinballs in a flipper machine. Or we can design the very same technical system in such a way that bike couriers benefit from a free and sporty job in which they have control over their own time, find a sense of community and remain private. We can build telemedicine platforms in such a way that they are no more than a cheap dial-in service – dehumanized, convenient and supported by a half-baked self-service AI. Or we can build such a platform as a cooperative online environment, helping remote patients and practitioners to benefit from global health knowledge expertise.

When values are explored with the Value-Based Engineering approach early on in an innovation project, what emerges is a whole new value story or likely value proposition reachable with the technology. The dormant value potential inherent in the technical capabilities becomes apparent – value potentials that are social and human-friendly and go beyond the narrow, profit-driven goals of efficiency, productivity or monetary gain pursued by the majority of IT innovation efforts these days. The case studies conducted in cooperation with my Ph.D. student Kathrin Bednar have opened my eyes to the fact that we could indeed create a better world with technology if we wished to. But will we do that? Will anyone care to use Value-Based Engineering, seeing that it takes more time and effort to develop a system with care? At least this book, together with IEEE Std 7000™, gives tech companies a clear, transparent, reviewed and sound methodology on how to go about such an undertaking.

This book guides its users on how to use IEEE Std 7000™ and how to interpret many of the standard's terms. It provides project teams with practical forms to fill when traceably deriving system features and requirements from value principles. But it also encourages the philosophically interested reader to understand what values are. Unlike any other contemporary work on values in technology (that I know of), it contains what philosophers call an "ontology" – an ontology of the value construct that allows the users of this book to discern value dispositions in systems, from value qualities perceived by humans to those absolute values we find, for instance, in laws or human rights agreements (see Chapter 3). This ontology was derived from Max Scheler's *Material Value Ethics*, a philosophical masterpiece ignored and underestimated by many contemporary value ethicists. I am particularly proud of this value ontology and the corresponding value vocabulary, which I could then apply in large part to bolster the standardization effort. I believe that without a sound definition and understanding of the value phenomenon itself, the IEEE Std 7000™ effort would have remained unsatisfactory.

I am especially grateful to my husband Johannes Hoff, who as a professor of philosophy and theology accompanied and continuously challenged me in this endeavor to develop a proper value ontology for technology design – an endeavor that would ordinarily be pursued only by "hardcore" philosophers and hardly IS scholars like myself! I am also grateful to Lee Barford, a mathematician, theologian and quantum physics veteran in Silicon Valley who is co-responsible for much of the ethical details in IEEE Std 7000™ that are now also embedded in this book. I want to thank Lewis Gray, a consultant on software system development based in Washington, who taught me more about IT requirements engineering than I could learn from any textbook in the world. Equally, I want to thank Ruth Lewis, an Australian IT consultant with whom I co-developed the EVR-construct (Ethical Value Requirements) key to Value-Based Engineering, and who opened my eyes to the interfaces between Value-Based Engineering and design thinking. I am indebted to Annette Reilly, Technical Editor of IEEE Std 7000; Ali Hessami, Chair of the P7000 Working Group (WG); and Gisele Waters and Alexander Novotny who patiently endured the IEEE Std 7000^{TM} project for as long as it lasted, leading it, improving it and making it what it has become. Without them, this book would not exist either. Ruth Lewis, Lewis Gray, Ali Hessami and Alexander Novotny have also reviewed various sections of this book and gave me invaluable feedback. I want to thank my Ph.D. students Till Winkler and Kathrin Bednar. Both helped me grasp the complexity of value clusters. Till also participated in the P7000 project and was a wonderful peer, equally reviewing parts of this book. Ross Ludlam continuously helped to polish and improve the language as my editor. Finally, I am deeply indebted to Marie Therese Sekwenz, who illustrated this book for me. My idea to develop drawings for almost every page was inspired by old biblical scriptures like the Lutheran Psalter, which always had drawings accompanying the text. The drawings served the purpose of helping to better commit the content to memory. Indeed, it would be possible for a skilled Value-Based Engineering lecturer to teach the material of this book based on the figures alone. Without Marie Therese's talent, this wonderful visual format would not have been possible.

Last but not least, I want to thank the 154 experts who were at some point involved in the making of IEEE Std 7000^{TM}, 34 of which were WG members. Seventy-seven experts balloted for the standard's publication in 2021 (93% acceptance rate). Without them and the generous support of the IEEE Standards Association (in particular, Konstantinos Karachalios) and IEEE Computer Society, neither the standard nor this book would exist.

At this point I must also make a disclaimer: In this book I cite extensively from IEEE Std 7000^{TM}, but I want it to be noted that IEEE standards like IEEE Std 7000^{TM} are the result of group work and consensus-building. Many small decisions are taken, and not all of these are shared unanimously. The interpretations and views expressed in this book solely represent my view on how IEEE Std 7000^{TM} can be put into practice. I do not necessarily represent the position of either the IEEE Std 7000^{TM} Working Group, IEEE or the IEEE Standards Association. Where I have felt that IEEE Std 7000^{TM} does not recommend what I think should be done or has a suboptimal definition of terms, I have made this explicit here, and also give reasons for my own divergent view where required.

Contents

Preface —— V

Chapter 1
Introduction —— 1
    Technological progress: GDP or wellbeing? —— 1
    IT gardens versus IT deserts —— 4
    What is Value-Based Engineering? —— 8
    What to expect from Value-Based Engineering? —— 9
    Challenges of profit-driven corporate innovation culture —— 11

Chapter 2
The 10 principles of Value-Based Engineering —— 15
    Value-Based Engineering is not bowling alone —— 17
    Principle 1: Ecosystem Responsibility —— 19
    Principle 2: Willingness to Renounce Investment —— 21
    Principle 3: Stakeholder Inclusiveness —— 22
    Principle 4: Use Moral Philosophies for Value Elicitation —— 24
    Principle 5: Context Sensitivity —— 27
    Principle 6: Respect for Regional Laws and International Agreements —— 29
    Principle 7: Leadership Engagement —— 32
    Principle 8: Transparency of the Value Mission —— 33
    Principle 9: Understanding Values in Depth —— 37
    Principle 10: Using Risk-Analysis for System Requirements Elicitation —— 38
    Ethically aligned design needs iterations and adjustment —— 39
    Check questions —— 39

Chapter 3
What values are —— 40
    Towards a definition of values —— 40
    Why is a rigorous value definition important? —— 43
    The process of valuation —— 46
    Technical valuation is not a matter of personal taste —— 47
    The importance of experience in value quality judgments —— 49
    Towards a three-layered value ontology —— 51
    The contextual meaning of values —— 54
    Conceptual value analysis —— 55
    Constitution of goodness through positive & negative values —— 58
    Check questions —— 59

Chapter 4
Value-Based Engineering phase 1: concept and context exploration —— 60
    SOI context exploration —— 64
    System-of-System analysis —— 66
    Partner analysis —— 71
    Challenges of concept and context exploration —— 73
    Stakeholder identification and roles —— 75
    Feasibility analysis —— 78
    Consequences of context and concept exploration —— 81
    Check questions —— 81

Chapter 5
Value-Based Engineering phase 2: value exploration —— 82
    Including stakeholder representatives —— 84
    Appointing Value Leads —— 85
    The philosophical foundations of value elicitation —— 87
    Practical issues in value elicitation —— 97
    Value clustering —— 100
    Prioritizing value clusters —— 102
    Resolving value trade-offs —— 107
    Conceptual analysis of core values —— 109
    Check Questions —— 110

Chapter 6
Value-Based Engineering phase 3: ethically aligned design —— 111
    Ethical Value Requirements —— 111
    Multiple paths of ethically aligned design —— 114
    Standard risk-based technical design —— 114
    High-risk-based technical design —— 122
    Check questions —— 132

Chapter 7
Transparency and information management —— 133

Chapter 8
Epilogue: dormant values versus gadgetism —— 136
    Pure-will innovation —— 138
    Real-value innovations —— 142
    Schoolbook innovation management versus real-value innovation —— 143
    Dormant values in disruptive innovation —— 146
    Real values versus needs —— 149

 Who are lead users? —— **150**
 Performative acts in corporate innovation —— **152**
 Summing up —— **153**

Appendix 1 Case study: the rate your teacher app —— 155

Appendix 2 Notes on the visuals used in this book —— 157

References —— 163

Endnotes —— 169

Abbreviations —— 179

Index —— 181

Raja Chatila (Professor, Sorbonne University, Paris)
"There is a wealth of books and publications about technology ethics, or recommendations for trustworthy AI systems. Recently, the IEEE published the "Standard Model Process for Addressing Ethical Concerns during System Design" (the IEEE Std. 7000-2021) under Sarah Spiekermann's leadership. But what are the foundations of ethical design? What are values, and what is the difference with principles? Above all, HOW to concretely develop a technological system that is aligned with humanistic values? While the European AI Act is still in the womb, Sarah Spiekermann united her knowledge, experience and talent together in this brilliant and accessible book which provides a comprehensive manual, bridging the gap between theory and practice. It will be the precious companion of every engineer and every manager, to understand how to build ethically aligned technologies, but also a resource to empower the interested citizen."

Lee Barford (Fellow at Keysight Technologies, computer scientist, theologian, Silicon Valley)
"Others have philosophized about applying ethical and technical wisdom to yield innovation that benefits humanity. This book brings together the understanding of human values, business leadership, and engineering and software processes needed to bring about ethical innovation in practice."

Lewis Gray (Software System Development Consultant, Washington)
"I sincerely believe that this book has the potential to revolutionize the systems development industry as compliant projects create attractive, ethical "canaries" in the nonethical (often unethical) "coal mine" of the systems development marketplace."

Ali Hessami (CEO of Vega Systems and former Chair of the IEEE 7000™ standardization project, London)
"Autonomy and algorithmic learning are defining features of key emerging technologies with pervasive use and wide-ranging social impact. Consideration of human values in the design of these technologies is addressed in the landmark IEEE 7000™:2021 standard, as the forerunner of Value-Based Engineering practice. The training underpinned by this comprehensive handbook will generate requisite insights and empower technologists and principal stakeholders to incorporate, foster and where appropriate, protect stakeholder values in their technological artefacts and services. As the chair and technical editor of IEEE7000™ standard, I commend this foundational training as the most insightful programme for Value-Based Engineering."

(Yvonne Hofstetter, Lawyer, Professor h. c. for Digitization and Society, CEO of 21strategies, Munich)
"Sarah Spiekermann belongs to the most significant voices of our times when it comes to Value-based Engineering. Her work focuses on development and facilitation of philosophical knowledge for system engineers. Spiekermann succeeds in a unique way in bringing natural sciences and technology into a long overdue concordance with the humanities, an effort aiming to make our highly technology-focused century more humane again. Her work inspires an unusually creative realization of positive machines, which strive for more than attaining the economic value of modern technical systems. In her book "Value-based Engineering" she takes the reader by the hand and guides her through theories of values and ethics, and also uses rich imagery to introduce her to those tools which help to ensure that technical systems of the future bring ever more joy and VALUE-add in the best meaning of the word."

Konstantinos Karachalios (Managing Director of the IEEE Standards Association, New York)
"What started as an exploration of thinking on computer ethics in Sarah Spiekermann's "Ethical IT Innovation: A Value-Based System Design Approach" has now culminated in this very remarkable practical guide to build ethical technology platforms through Value-Based Engineering. Between these two milestones resides the development of the IEEE Standard 7000, "Model Process for addressing Ethical Concerns during System Design", that was inspired and initiated by Sarah Spiekermann, and which constitutes one of the foundational elements of Sarah's new book. Beyond its inherent unique value for IT systems designers, this book is also a testimony about a passionate journey leading from high-level principles and values to concrete and pragmatic guidance and practice. Such a journey affords a very intense personal commitment in the messy waters of rules-based, collective consensus building over many years, not an evident call for a teaching academic. But a compelling one for a person with a positive vision for humanity, a love for ethics and philosophy as well as a passion for good governance of technology. We were really lucky and privileged to have joined forces with Sarah to advance toward this goal."

Wolfgang Koch, Chief scientist at Fraunhofer FKIE, Bonn
"For knowledge itself is power ... " – Francis Bacon's famous statement marks the beginning of the modern project. For all who wish to harvest technological power for the benefit of humanity, Sarah Spiekermann's guide to the galaxy of Value-Based Engineering is a "must read". In Age of AI, she makes the new IEEE 7000™-2021 process model transparent which turns ethically guiding knowledge of the human nature and its "ecology" into sober systems engineering principles.

Paul Nemitz (Principle Advisor, European Commission, Directorate-General for Justice and Consumers, Brussels)
"Sarah Spiekerman paves the way for engineers to combine imagination, innovation and societal responsibility commensurate with their new and exponentially increasing power in the age of artificial intelligence. Individual engineers will be able to discharge their new responsibilities only if they personally engage their genuine human ability of critical reflection and deliberation with others in all stages of their work. This book will be a starting point in this endeavour which will eventually have to lead to an emancipation of engineers from the domination by profit objectives to a culture of engineers who as individuals in all their doing act as responsible citizens who contribute to ecological and social sustainability, the good functioning of democracy and the rule of law and the delivery of fundamental rights and civil liberties and who discharge this duty even in adverse circumstances and where necessary against the pressure of profit objectives of their employers or customers."

Chapter 1
Introduction

The modern age is marked by a deep-rooted belief in the merits of technological innovation for human wellbeing and evolution. Ever since the thirteenth century, when the word *innovare* was first used by the German monk Albertus Magnus, humanity has been driven by the belief that new tools and new production and building techniques foster human advance. "No empire, no religion, no heavenly body can have a more fundamental influence on human conduct . . . than these mechanical inventions," said Francis Bacon (1561–1626) at the dawn of the current Western civilization, expressing a maxim that gained tremendous global influence. New weaponry, machinery, transportation and health infrastructure, clocks, digitization and, lately, quantum computing have allowed humanity to accumulate such a degree of wealth and health that any critical doubting of the innovation culture is met with pushback by the established market forces. In contrast, newness is regarded as *automatic goodness*, and many of the world's major problems, mostly caused by technology, are now, we hope, equally able to be fixed by it.

Against this historical and cultural background, the digital transformation of production and society is embraced with little doubt as to its merits. The idea to digitally permeate everything that lives seems like natural advance. But do increased levels of production, government and household automation as well as digital mediation of social processes really lead to the almost linear form of continuous progress expected by our economic and political establishment? This book doubts that this is *automatically* the case. Instead, it starts from the hypothesis that technological transformation leads to progress, only if it is actively shaped to create positive human and social value. The digital fabric is like fire or electric energy: it bears great positive value potential for humanity. However, it unfolds this service to humanity only if it is used wisely. Otherwise, it can also scorch the cultural, spiritual and economic soil that it set out to enrich. Humanity can just as much stumble into a stage of unexpected regress with digitization, as fire can burn down a house. Only an active shaping of digital services for positive value creation (Figure 1.1) can turn the wheel in humanity's favor. This book is a guide on how this can be done.

Technological progress: GDP or wellbeing?

Where do we stand today in terms of digital progress? So far, value creation is, often, primarily equated with monetary value. Economic systems worldwide and the theories catering to them equate the wealth of nations with financial utility. Gross domestic product (GDP) represents the monetary value of what is produced in a nation. And, from this monetary "value" perspective, digitalization has had extremely beneficial

2 — Chapter 1 Introduction

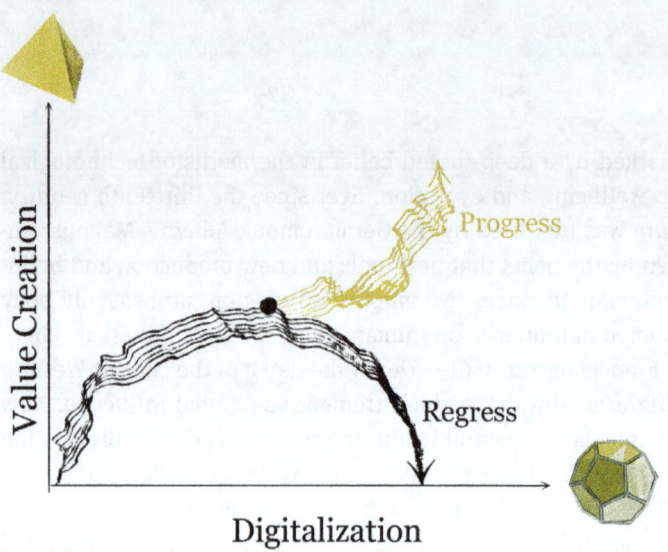

Figure 1.1: Progress or regress through digitalization?

effects on all economies that manufacture goods and provide services. Digital automation allows companies to realize economies of scale, save significant production cost and increase work as well as capital productivity. Through corporate digital networks, especially ERP systems (like SAP), global markets can be served much more effectively out of one (instead of several costly) corporate headquarter, leading to positive GDP effects in all those countries where headquarters are based. At the corporate level, value in terms of profit soars when labor cost can be reduced or when costly office-rents are saved with home-office. Digital decision support systems speed up transactions and, thereby, the volume of what can be traded and managed. In this regard, the value creation curve triggered by digitalization (as depicted in Figure 1.1) has seen positive growth in the past four decades.

That said, the idea that "value" can be reduced in economics to a monetizable unit is increasingly being contested. The Gross National Happiness Index used by the state of Bhutan, for instance, has been an early frontrunner in the rising global awareness that GDP or monetary indices of production are not sufficient indicators for reflecting the *value creation* of nations or understanding the value of living conditions of individuals. In contrast, it has become clear that the wellbeing of people and the ability to sustain earthly resources is at least as important a value, if not more than monetary value.[1] A shift has begun towards embracing politically and economically, what has been philosophically clear for a long time: that value creation is about the myriad qualities that foster human and environmental wellbeing, of which money is only one. Figure 1.1 should be understood in this way. In the future, positive progress through digitalization can only be achieved, if aggregate human and social value in terms of wellbeing is borne by digitalization. If there is no increase in aggregate value, regress may equally set in; so where do we stand today, if we view value creation through this prism?

To answer this question, wellbeing must be better defined than is currently the case. The World Happiness Report published by the United Nations Sustainable Development Solutions Network (SDSN) on an annual basis is a first step in this direction. It incorporates various values that are both perceived by human beings and borne by forms of government and national infrastructure, including freedom, generosity, health, social support, income and trustworthy governance (SDSN, 2022). But strictly speaking, the aggregate of these select values can only give a broad notion of what is really driving wellbeing at an individual level, since values and, indeed, wellbeing are deeply contextual. Thus, a person can be happy even if she is not free; or she may be very unhappy even if she is healthy and living in a salubrious environment. And, what about values such as knowledge, security or dignity? Are these not important for wellbeing? These critical questions do not aim to demean the effort of the United Nations SDSN. In contrast, the report is an important tool in raising governments' awareness for the kind of progress the world should embrace – one that is defined through human and social values. However, a robust definition of wellbeing should go further and recognize the contextual dependency of personal value perception. Life is good on a day-to-day basis, only if the contextual conditions in which human beings normally live are imbued with meaning and purpose; that is allowing for "valueception" in what one is doing, especially in one's community. At the same time, living conditions should not impede flourishing by bearing too many "negative value qualities," such as a lack of freedom to move, a lack of food or health support. What is encouraging or hampering wellbeing can depend on myriad positive or negative value qualities at a time, differing from one context to another. Thus, any shortlist or metric for value creation or wellbeing can only be understood as a snapshot and excerpt of a much more complex reality.

This book recognizes the complex contextual nature of values and value creation. It does not recommend that companies and governments concentrate only on a few value principles that they implement and measure across contexts or technologies. This kind of "solutionism" is not enough for understanding of values or ethical engineering covered by this book. Instead, this book serves as a guide to help companies and governmental institutions to envision the wide range of positive and negative value potentials or phenomena of desirability (and undesirability) associated with a new technology product or service they wish to launch. And, for each of these contexts and product technologies it advises innovation teams to anticipate, explore and understand the myriad value effects and consequences for wellbeing.

Technology projects in this book are likened to garden development projects (Figure 1.2): Each garden is different and changing, depending on climate, soil, landscape and purpose. Each garden has its own challenges. And, the only progress that can be achieved is that the existing value potentials of the ecosystem are intelligently understood and taken into account, so that a good and beautiful place can unfold, despite all remaining contingencies. A garden changes its value during season and over time. And, gardeners – just like engineers – need to adapt. Equipped with this understanding, innovation teams (gardeners) are encouraged to define context- and technology-specific "ethical value requirements (EVRs)," which are the context-sensitive criteria for a mindful and value-based IT system (or garden) design and development.

Figure 1.2: VBE seeks to create IT gardens.

Progress or regress at the point of inflection in Figure 1.2 is determined by the degree to which innovation efforts are able to create IT gardens for humanity versus efficient, but hardly sustainable and livable machine–deserts (Figure 1.3).

IT gardens versus IT deserts

At the time of writing this book, a debate has emerged as to whether IT innovation and digitalization are actually leading to gardens or to human, social and environmental deserts. On one side, liberal and radical post-humanists as well as many industry thought leaders herald the dawn of a "Fourth Revolution" through technology (Harari, 2017; Schwab, 2017). On the other side, critical voices have started to undermine the formerly steadfast belief in human progress through digital innovation.

Figure 1.3: Non value-based IT deserts.

Despite all production efficiencies and positive GDP effects, increased work flexibility, global corporate control, accessibility of people and cultural goods (e.g., digital music, films, arts, games), veterans of the early days of Silicon Valley's powerful service oligopoly have started to turn away from their former employers and publicly warn of the "social dilemma" created by digital technologies (Orlowski, 2021). Humans are "degraded," they say, because online services are, for the most part, pursuing business models that Shoshana Zuboff has identified as "surveillance capitalism"(Zuboff, 2019). They are built on the trading of personal data, and the capitalization of human attention and manipulability. Sixteen million people spend time in virtual worlds on a daily basis in the USA alone – a trend that is predicted to rise further.[2] People use mobile phones for over three hours a day[3], on average, and fall into habits of self-interruption, impeding their ability to feel positive flow, to learn, concentrate and actually *see* the real world around them. Many are trapped in online social network bubbles and echo chambers that manipulate them into strange behaviors and kindle dangerous beliefs – from fearing that all birds are drones spying on people to more serious fake news that may drive social media users to embrace extremist positions, for instance, in elections or in matters of political conflict. The end of democracy has been voiced as a concern due to these developments, a threat that overshadows less visible, but at least, as problematic developments at the psychological level: namely, young people seem to be growing into what psychiatrists have started to call "e-personalities" that lack prudence, patience and humbleness in dealing with the world (Aboujaoude, 2012) (Figure 1.4). The Internet has taught a whole generation that everything and everyone – knowledge, people and service – are always available to them, effortlessly, and are only a click away.

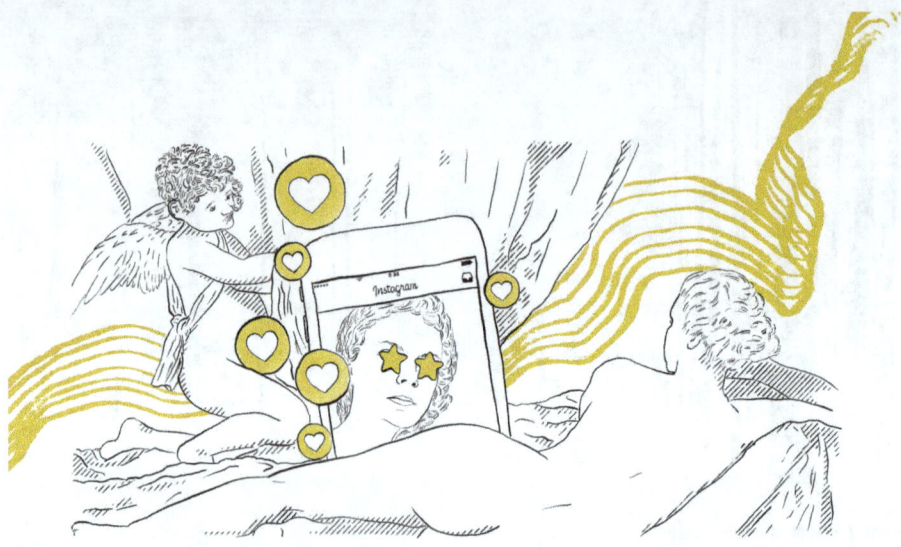

Figure 1.4: Flourishing e-personalities.

But this reliance of the digital generation on the supportive and, presumably, secure and trustworthy frame of technology is increasingly questionable, if not outright dangerous. As software starts to replace and regulate hardware components, systems have become more opaque and, sometimes, less reliable and predictable, especially when they are rushed to market. This became painfully clear in the case of the Boeing 737 Max plane crashes, where the pilots did not understand why the autopilot was sending the aircraft into a nosedive. The wider use of Artificial Intelligence(AI) has uncovered the challenges associated with the use of historic, sometimes low-quality data in decision support systems and workflows. Software predictions and recommendations are often imprecise, biased and error-prone, even if the code is of good quality. But code itself is also on trial. Today's software programs are often an accumulation of preconfigured code components and, more or less, black-boxed algorithms, which are cobbled together into a new system. While this way of building software programs implies huge time and money savings (since developers do not have to develop a system from scratch), it also means less control over a system and high dependence on the availability and quality of external services, not to speak of the complexity of service agreements and dependencies.

All this implies that at the time or writing this book, IT systems are less secure and, often, less trustworthy than the traditional analog and mostly mechanical machinery, the reliability of which has taught humanity to believe that machines do not fail, unless they are broken. Every year, billions of data records are exposed, often due to misconfigured databases and system components.[4] In 2020, hackers worldwide attacked over 30,000 websites, every day; every 40 s, a new cyberattack was initiated, costing, on average, US$4 million[5] per attack.[6] This vulnerability comes on top of error-proneness of digital systems. It is often forgotten that code is hardly ever perfect and can lead to tens of thousands of mistakes in complex systems like autonomous cars.[7] AI used in the US justice system, for instance, was found to condemn people to jail on the basis of software biases (Hao, 2019). Finally, the unchallenged timing and error-proneness of corporate IT systems, combined with the constant availability of the network, influences peoples' health (Figure 1.5). The number of employees who suffer from fatigue and emotional exhaustion seems to be rising and often tips over 25% of the general working population, depending on country and assessment method (Aumayr-Pintar, Cerf, & Parent-Thirion, 2018; Shanafelt et al., 2019). To make a long story short: IT deserts and regress, rather than IT gardens, seem to be on the rise.

Figure 1.5: IT systems undermining peoples' mental and physical health.

A core hypothesis of this book is, however, that a lot of these negative value effects could be avoided *if* IT systems were better designed and *if* the innovation processes accompanying system design were focused on human and social values. What is needed is a new culture of IT innovation, system development and system testing that is less focused on money, efficiency and speed. A culture that is diligent, controlled, open, transparent, quality-oriented and, most importantly, accompanied by processes of foresight and care. Such processes of foresight and care can anticipate many potentially negative value effects caused by IT systems and develop strategies to mitigate them, before launching a product. However, value-based processes of care also bring about the question of what a system is good for, in the first place, why it should be there, what human and social values it should actualize and, thereby, how it is meant to contribute to the wellbeing of society. A culture of care can be established in system development and innovation – one based on values. This book is a guide on how to potentially grow and nourish such a culture in companies with Value-Based Engineering.

What is Value-Based Engineering?

Value-Based Engineering (VBE) is a new, profoundly visionary and wellbeing-driven way to live and breathe IT innovation. It starts off at the earliest possible moment of business mission definition and accompanies innovation teams through various processes, all the way through to system requirements. At its core is the world's first model process for addressing ethical concerns during system design, IEEE 7000™. IEEE 7000™, in publication with ISO as ISO/IEC/IEEE 24748-7000 at the time of writing this book, has been developed with the help of over 35 standardization workgroup members, over the course of five years (2016–2021) and reviewed by 95 international experts. The workgroup replied to over a 1,000 improvement recommendations from these experts. As a result of this inclusive and open collaboration, many of the world's most proven approaches to value-driven system design could be considered, including stakeholder involvement, value conceptualization and risk-based design.

The IEEE 7000™ standard details these existing and new best practices in five engineering process sections needed to build a value-based system. In this book, these are condensed into three chapters listed below, and Transparency Management (described in Section 11 of IEEE 7000™) is covered through five reporting forms (Figures 5.10, 5.12, 6.2, 6.5 and 6.9) that have been developed and tested to accompany a VBE project:
1. Concept of operation and context exploration
    (Section 7 of IEEE 7000™) (Chapter 4)
2. Ethical values elicitation and prioritization
    (Section 8 of IEEE 7000™) (Chapter 5)
3. Ethically aligned design
    (Sections 9 & 10 of IEEE 7000™) (Chapter 6)

The first block of work that is needed in a VBE project is to literally prepare the *soil* for a new System of Interest (hereafter "SOI"). This is why a brown cube is, hereafter, used as a symbol for this phase of system engineering (Figure 1.6). The brown bare earth needs to be studied: Project teams seek to understand the SOI's use case, the context where it is supposed to be used, the stakeholders involved, the various system components planned to enable it, the data flows and the partner structure.

Once this is understood and a first Concept of Operations (hereafter "ConOps") is sketched out for the SOI, value elicitation can happen. It uses established moral philosophies or spiritual traditions to uncover the many dormant value potentials inherent in a new SOI. After these are collected from stakeholders, they are conceptually analyzed and prioritized to inform subsequent system design. As Figure 1.6 shows, values are symbolized through tetrahedra, because just like tetrahedra, they have many qualities (sides), only some of which are directly visible to stakeholders (for more detail, see Chapter 3 and Annex 2).

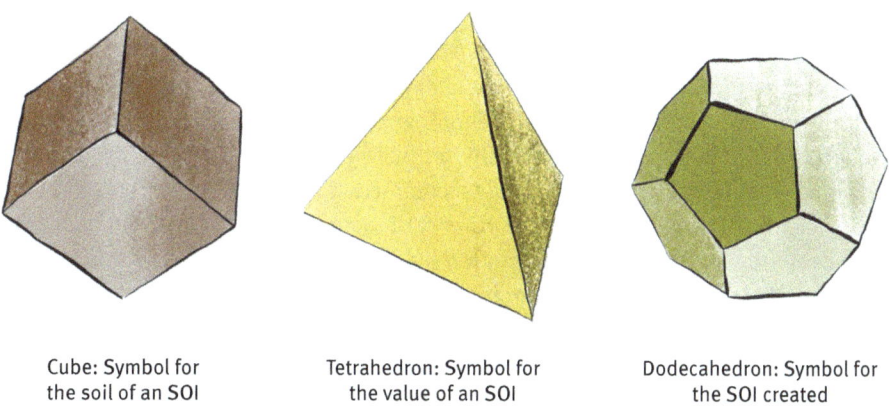

Cube: Symbol for the soil of an SOI

Tetrahedron: Symbol for the value of an SOI

Dodecahedron: Symbol for the SOI created

Figure 1.6: Platonic solids symbolize VBE constructs.

The final block of work needed for VBE is the actual design and specification of the SOI. This is done by deriving so-called "Ethical Value Requirements" from the core values identified earlier and translating these with risk analysis into concrete system features. Since system features integrate many value ideas (tetrahedra), a dodecahedron is chosen to represent the final SOIs in VBE (Figure 1.6).

What to expect from Value-Based Engineering?

What can companies expect from a VBE project? At the time of writing this book, companies engaging in ethical or value-oriented work seem to be mainly driven by the desire to comply with legal expectations on their systems, such as the EU's AI Act or the General Data Protection Regulation (GDPR). While it is nice to think about corporate values in abstract, earnest financial investment in value-based system design only seems to happen when legal sanctions are looming or when ESG criteria need to be fulfilled to qualify for the attraction of sustainable investment money. Certainly, companies that engage in VBE can expect that, by following the processes described in this book and applying the IEEE 7000™ standard, they will be in a very good position to comply with the law. VBE processes force innovation teams to deeply scrutinize the social and governance issues of a new SOI, and potential privacy issues are detected and prioritized.

That said, VBE is not at all only about reaching legal compliance, because acting in a legally defensible way is not the same as acting ethically. Building systems ethically means to build them in a good and right way for a community of human beings (not "users") and to strive for wise and worthy ends in the deployment of a system. This is a much bigger vision. It is the vision to build technology for humanity and not for profit maximization.

A concrete VBE case study that may illustrate this was conducted on the technology platform underlying a food delivery service, such as Uber Eats or Foodora: The initial goal of this case study was to develop a traditional product roadmap as it is typically used in today's corporate IT planning: Agile processes are used, inspired by available and new technical functionality to build technology for profit-driven business models. Analysts of the food service identified many opportunities to design the technology: An AI could optimize the number of routes per courier and minimize couriers' break times, telling them, through a voice-interface, what street-turns to take. An AI could also recommend restaurants to food purchasers that offered the highest profit margins for the courier service. The platform could survey and analyze couriers' performance, handing delivery jobs preferably to those that were fastest and most flexibly available. Moreover, food consumption data as well as platform-use intensity might be an interesting social-psychological data point to sell to data markets. All of these realistic service features share one central value they optimize – that is, the monetary financial value to the corporate shareholders of this service, a value driven by a few other instrumental values such as efficiency and productivity. What was largely ignored by the product road mappers was that all these system features bear negative value externalities, such as a loss of freedom and autonomy, as well as privacy on the side of the couriers. Stress and health issues could equally ensue when bike couriers are managed by an AI, like pinballs in a flipper machine. This ignorance of negative value qualities created by a technology-driven road mapping practice completely disappeared, when applying the VBE value elicitation processes described in this book. In reflecting about the stakeholder values impacted by such a service, challenging potential value and virtue harms as well as duties and positive value opportunities were identified, and a whole new service story and IT vision emerged. Each case study analyst not only saw, on average, ten negative value risks looming to undermine investors' long-term enthusiasm about such a service, but also had, on average, 13 product ideas driving the positive value potential of such a service (Bednar & Spiekermann, 2022). They thought about how gamification might be used to support the joy of couriers delivering food; how giving them control and autonomy over route selection and privacy and treating couriers on equal terms would benefit their health, comfort and satisfaction, in doing the delivery jobs. They came up with new business models where the platform could support customers' diet projects and where couriers could couple their delivery job with that of a health coach – not only supplying the food but supporting customers to maintain diets. They also thought about the loneliness of couriers that could be mitigated by enhancing the platform with courier-community functionality, sharing courier hangout places to overcome idle time. Even the cooks were recognized as stakeholders: their joy and motivation might be fostered, if a positive feedback-channel to customers were implemented. All these visions do not impede the earning of money with the service, but they balance the bottom line with the human and social impacts of the service.

What this case study illustrates is that by employing the value-elicitation processes described in this book, a whole new innovation culture is born; an innovation culture that is living and breathing the idea to build technology for humanity, for individual and social wellbeing. The delivery platform case was just one of several case studies. All we did witnessed the same phenomenon: as soon as innovation teams are asked to actively reflect on human and social values and derive system features from them, technology is put to different uses. It is put in the service of humanity (for more information on case studies conducted, see Bednar & Spiekermann, 2022).

Challenges of profit-driven corporate innovation culture

Given the amount of work and care that needs to be put into a value-based system, it becomes clear that companies that want to engage in VBE must probably rethink and reconfigure parts of their innovation and engineering processes and culture. This is not an easy endeavor in the context of today's mainstream corporate milieu, where short-term shareholder value trumps most other possible goals of a corporation (where Mammon rules (Figure 1.7)). Profit maximization potential, satisfactory performance and time-to-delivery are, often, the core goals of both the management units and the system developers, when a new product or service is brought to market. These business priorities often shape the processes, decisions, tools and habits that are used to build IT systems, at least at the time of writing this book.

As described in the epilogue to this book (as well as in Chapter 8), it is, nowadays, often not a value vision or a deep-seated human need that kicks off innovation, but the availability of a new IT functionality. Every year, around three million new patents wait to be monetized,[8] and with these, come the IT industry's hype cycles, such as Big Data, AI or the Metaverse, which call for ever new investment. The social reality that is supposed to absorb all these technologies is not really cared for, as a force in its own right. Instead, a kind of faceless "user" market is assumed, which the IT industry presumes to simply shape and "design," as it wishes. Through marketing efforts,

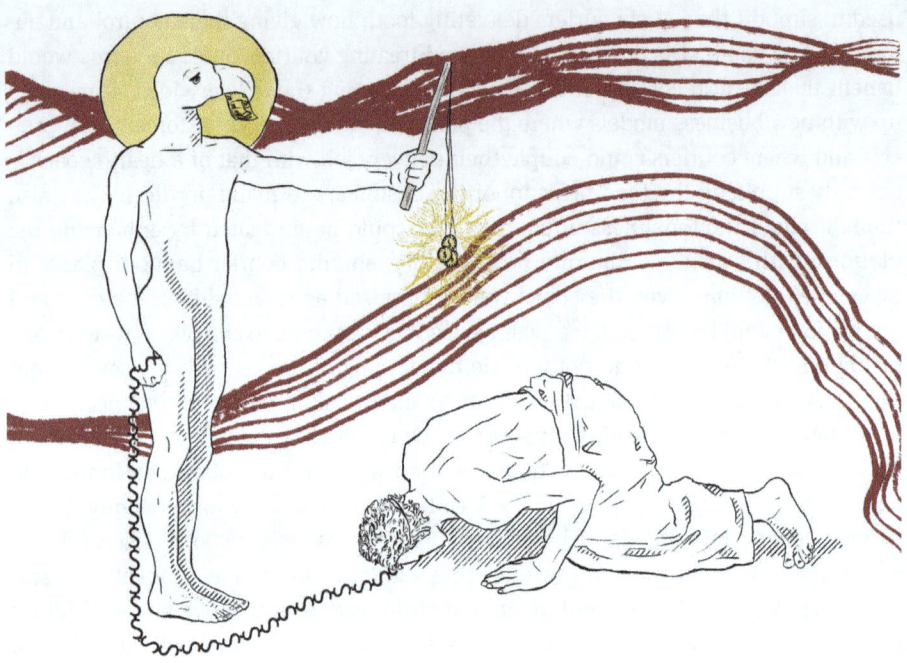

Figure 1.7: Mammon ruling over manager decisions.

freemium models or simply by corporate order, innovation is poured over markets, no matter how void they are of any real customer value. The fact that more than half of IT innovation efforts thereby fail,[9] and also half of all US start-ups don't survive more than five years is ignored,[10] and this waste seems to be accepted as collateral damage and investor risk. Yet, the waste should not be underestimated as to its real cost, because people involved in these processes lose valuable years of their life, and users confronted with half-baked services waste time, nerves and money as well. Wasteful innovation efforts breed a yet-unmeasured amount of opportunity cost at the aggregate societal level. VBE can reduce this waste, as it anticipates many of the value harms that can undermine the success of an innovation project.

However, even if IT innovations are successful in the market, a question remains: What kind of value do they actually contribute to human and social progress? The traditional economic argument would be that as long as something new generates income, its existence and continuation are justified and good – after all, the innovation contributes to the growth of GDP. Customers chose and paid for the new product, signaling that the innovation is effective in addressing a human or social need. GDP and social value can be assumed to overlap, to some extent. However, IT services' income is not necessarily, any longer, created through primary digital product or service sales that signal a market appreciation. In the digital economy, primary services are often for free, and income is generated through secondary sources, such as the sharing of customer data or selling of customers' attention (Figure 1.8). In "Surveillance Capitalism," as the digital economy has been critically termed (Zuboff, 2019), market appreciation and corporate income generation are often decoupled. Technical functionality is put in the service of secondary income generation, which often creates negative value externalities, such as the loss of customers' privacy, dignity, freedom, creativity, community or identity (Figure 1.8). With these developments, GDP progress does not automatically mean social progress any more.

VBE goes back to the drawing board, here, in that it seeks to prohibit negative externalities. Furthermore, it supports companies in identifying higher human and social values that might foster the attractiveness of the primary technical innovation. A value-driven culture is created that revives the opportunity to make money from primary digital services.

Figure 1.8: Surveillance capitalism promises revenue from value-poor services.

Whether a value-based culture can be created depends, however, on the degree of profit focus and engineering culture a company has. The strong profit focus influences the processes by which technology is engineered today: a quick and agile way, speeding up development time to a minimum, accompanied by lean requirements' engineering with little documentation, a lot of outsourcing as well as a trial-and-error approach to market (release early, release often) certainly drive the corporate bottom line. However, this kind of fast system engineering is often accompanied by a lack of care, time and up-front planning – all of which are necessary for VBE. VBE promotes what seems to have fallen into oblivion: namely, that good requirements' engineering at the beginning of innovation projects is very important, before agile forms of development can actually start in earnest. As Laozi (604–531 BC) famously said (Figure 1.9):

"Only those who know the goal, find the way." If it is not clear what human and social values a system should foster besides monetary value, then the goal is not clear; the question of "why" a product or service should be developed is unclear. The system is void of human purpose. Yet, purpose is important for anyone who is involved in an innovation project. When people know that they are truly working for a good cause (purpose) and are about to develop a digital service contribution to society, then they are motivated to bring it to market, they are creative and they don't hastily quit their job for a better option. As VBE identifies purposes and ensures that IT products are value-rich, it gives back meaning to innovation teams and system developers.

That said, VBE will struggle in a milieu of harsh profit orientation, where time pressure and premature technical releases are the norm. At the time of writing this book, many IT development organizations suffer from a perceived lack of responsibility in their engineering units. Studies at the WU Institute for IS & Society have shown that while 90% of system engineers participating in a study on privacy and security engineering embraced the importance of privacy and security, almost 40% did not feel responsible for these values in the product. The core reason given for this lack of perceived responsibility is that organizations do not allow for enough time, training and autonomy to actually embrace such responsibility (Spiekermann, Korunovska, & Langheinrich, 2018). As a former Siemens engineer put it: "We see a rush towards system delivery that can lead to problematic ethical behavior, such as the rushing of engineering jobs, the difficulty to honor agreements, the lack of comprehensive and thorough evaluation of computer systems, a promotion of "fictionware" and the tendency to sweep a lack of quality under the rug" (Berenbach & Broy, 2009). VBE with IEEE 7000™ seeks to change this.

"Only those who know the goal, find the way."
(Lao Zi)

Figure 1.9: Lao Zi (571 BC).

Chapter 2
The 10 principles of Value-Based Engineering

In the aftermath of the September 11 terrorist attacks on the New York Twin Towers, most countries decided to ramp up security technologies, and in this vein, airports worldwide started to install full body scanners to screen passengers for weaponry or explosive materials, before boarding a plane. The goal was to strengthen airport security and flight safety. Several scanner alternatives came to the market (Figure 2.1), and it soon became clear that this technology could considerably harm passenger privacy by exposing people's intimate bodily details. For a short while, it seemed as if there was a trade-off to make between privacy and security. It turned out, though, that scanners built with privacy, by design, could resolve the issue. Presenting passengers as stick figures or schemata on security screens allows them to be scanned for security reasons, without exposing their figure or genitals. A privacy-sensitive technology version was born, which we now often use when traveling through airports. For the company (L3) that offered the privacy-friendly scanner, a significant competitive advantage was created. It was calculated that L3 could probably make over a billion euros in European turnover alone, if two of their scanners were purchased by every European airport (Spiekermann, 2012).

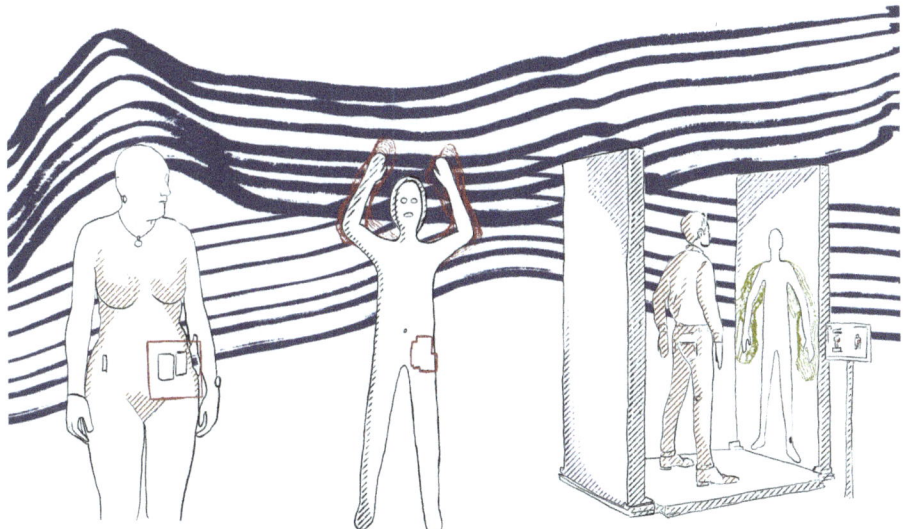

Figure 2.1: Technology evolution in airport body scanners.

However, have you ever been through such hands-up scanners yourself? Asked to raise your hands like a criminal? Legs apart, hands up? How did that feel? A value analysis would show that the bodily posture we have learned to associate with criminal conviction induces a negative feeling-state in us; a feeling of shameful unease. People rather feel as if they were suspected criminals being forced into a surrender position, instead of ordinary passengers. The negative value of default distrust is created, a perception of loss of dignity as well as discomfort. These value breaches give rise to such negative feelings that the privacy-sensitive design of the scanners alone has simply not been enough to corner the market, at least not in Europe. In fact, at least one European competitor grasped the opportunity to come up with a third solution, which offers passengers the possibility to keep their hands down. This is now a much more agreeable outcome for passengers, and walking through many European airports in 2021, it seems as if airport procurement is increasingly embracing this solution.

The airport scanner example has various takeaways:
- First, it shows that the Value-Based Engineering (VBE) of products can have tangible economic consequences, creating an important competitive advantage for those who envision and respect ethical consequences of their system design.
- Secondly, it shows that values that initially appear to be in an insurmountable trade-off, such as privacy and security in this case, can be overcome through good technology design.
- Third, it shows that values such as privacy and security, which have been well documented as tech-policy issues today, are not the end of the story when it comes to human-centric and socially acceptable technology. More values play a role, depending on the respective context (here: trust, dignity, comfort).

Note the term "engineering" in VBE: This book refers to "engineering" because providing ethically aligned security scanners is not just a "design" issue. It is also a technical and organizational engineering challenge. Even though the designer's sketch of a hands-down stick-figure-scanner is already a stroke of genius when compared to an exposing-nudity-scanner, it is not enough to build and operate good technology. What is also required is that the machine respects privacy in the background, works safely when being used and is dependable, so that its security scans can be trusted. This, again, means that the technical engineering of the scanners must be reliably executed in such a way that it lives up to these expectations. Who would be satisfied with a privacy-friendly screen image at the airport itself, when, at the same time, the full resolution nude picture version was being shared? Who would find it reasonable to use these scanners, if they did not do their job of truly detecting terrorists? And, who would want to be forced to use scanners that impact one's health due to radiation? These rhetorical questions show that the technical backend operations, data flows, organizational policies, operational testing and risk assessments are all required to make airport scanners work and to support their deployment from an ethical perspective. This goes far beyond design issues.[11]

Of course, one might wonder whether all these values – privacy, dignity, reliability and health – are not stating the obvious. Do passengers not take for granted that airport scanners and their operators work in a way that is reliable, safe, secure, and privacy-friendly? Unfortunately, at the time of writing, people's natural expectations of tech providers' respect for such obvious human values are often not met. In 2015, for instance, the news platform, *Politico*, reported that the US Transport Security Association failed to find fake explosives and weapons in 96% of covert tests of the scanners (Scholtes, 2015); so even the most essential functional value of this public system, its reliability and the ensuing security of passengers, was not provided.

The takeaway is that values end up being borne by systems, only when there is a rigorous engineering process. VBE provides for such processes. VBE, which is now, in large parts, standardized in the IEEE 7000™ Model Process for Ethical System Design, provides a structured and transparent method to ensure that organizations are aware of the full value spectrum impacting their stakeholders, so that they are then able to translate this into organizational processes as well as technical roadmaps (IEEE, 2021a).

Value-Based Engineering is not bowling alone

VBE is not a stand-alone practice that one organization alone can easily achieve. To provide IT services today, hardly any provider is an isolated greenfield entity, any more. IT systems often bring in a history. They are highly networked, if not interwoven with external web services (Figure 2.2). In VBE, as in many IT standards, we recognize that there is one "System of Interest" (SOI), but that this is embedded in a larger "System of Systems" (SOS). And, if an organization wants to respect values and ensure ethical conduct in a modern system, then this is only possible by choosing the right partners.

A simple illustration of an SOI is a webshop. A webshop will need to offer a digital payment function to its customers, but it is unlikely to have the competency to handle all the payment transactions by itself, in addition to its core business of selling a wide selection of products. Therefore, webshops typically delegate the handling of payments to a specialized credit card service like Mastercard or Alipay. An interface is created between the webshop and the digital payment service, and when a sale is agreed, the relevant purchase information is handed over from the webshop to the credit card service to do the billing and money collection. This is a good way for every system operator in the digital supply chain to concentrate on its own core competencies and to realize economies of scale in its own operations.

That said, what happens if a webshop provider – let's call him Peter – found out that all his customers' personal data, what they bought, at what price, at what volume, at what frequency, where they live, etc. would not only be used once by his credit card transaction partner to do the billing for him for a fee, but would

Figure 2.2: Responsibility for complex SOS computing networks.

continue to be used by the company for its own benefit? In a 2017 report on Corporate Surveillance in Everyday Life, Vienna activist Wolfie Christl reported how the company VISA, for example, "provided data on 14 billion purchase transactions to the data broker Oracle and combined it with demographic, financial, and other data in order to help companies better categorize and target consumers in the digital world" (p. 23 in Christl, 2017). Would Peter, the webshop provider, care if he learned that his credit card partner engaged in a similar data-sharing practice? If Peter wanted his webshop to protect the privacy of his customers, then yes, he probably would care. Just imagine Peter trading stuff like intimate toys or esoteric gadgets that none of his own customers would want to be publicly associated with. Peter would want to know for sure what his credit card partner is doing with his customers' data trails.

This is exactly where VBE comes in. Unlike most other approaches to ethical or value-sensitive system design, it always asks the question of ecosystem responsibility. An organization that wants to claim that it has built its system in line with its customers' values or with ethical principles in mind will always be at risk of disappointing, if it does not ensure that all its relevant partners are toeing the line. Therefore, the first principle of VBE is Ecosystem Responsibility: "Value-Based Engineering organizations embrace responsibility for their technical ecosystem. They abstain from partnerships or external services over which they have no control and which they cannot access."

Principle 1: Ecosystem Responsibility

Value-Based Engineering checks on AI service coupling

VBE with IEEE 7000™ is not the first approach to recognize the importance of ecosystem responsibility. Important ISO standards such as ISO/IEC 29101 (ISO/IEC, 2018), ISO/IEC/IEEE 15288 (ISO, 2015) or the European Data Protection Regulation (EU Parliament and the Council, 2016) have recognized how important ecosystem control is, for instance, for privacy reasons. It is vital to not only look at the data processed in one's own controlled IT environment, but also to look at the data exchange with other partners and what they are doing with the data.

Yet, those who think that data protection is the only value relevant in a responsible ecosystem are mistaken. Many values are at stake, when partners do not act in concert. Take the value of transparency. A system-of-interest provider might need to know how an interconnecting AI service calculates its results, before integrating these in her own service (Figure 2.3). If an external AI component is a black box and unable to explain how it achieves its calculations, then a responsible organization cannot integrate it in its own operations. Working in a value-based manner and with a sense of ethical responsibility, the organization would have to forgo the partnership.

Figure 2.3: AI services as part of an SOS computing network.

Let us clarify this issue with a real case from the university context. The goal of a university's admission office was to automate the processing of motivation letters of student applicants. It got thousands of cover letters each year; having those read and scrutinized by an AI seemed attractive. The university's AI project fed all of its application letters from a single year into an external text-analysis AI run by one of the world's leading AI providers. This external AI service returned a score on the level of motivation demonstrated by the student applicant in his or her motivation letter. Furthermore, it returned a calculation of the presumed personality of that student according to the so-called "Big Five" personality traits. When interviewed, the project lead had no clear idea of the algorithm design or logic employed by the external AI provider. He knew almost nothing about how the external service would calculate the scores (motivation, personality dimensions, etc.) The only thing he knew was that the external AI was trained on Twitter data. As a result, the AI's training data was from a context completely different from the one needed to evaluate student applications and, therefore, the decontextualized student scores were, probably, no more than senseless noise. No process was in place to evaluate whether the external AI scores would make any sense in the student application context. The idea to use the external AI was, consequently, abandoned. The example shows how important it is for an ethical organization like a university to know the exact details of an AI-operations partner. Only in this way can it exercise ecosystem responsibility. A responsible VBE organization would cease further cooperation with the AI provider or alternatively, in line with IEEE 7000™ (IEEE, 2021a), it would explore whether there is leeway to:
- cooperate with the AI provider on algorithm design
- co-determine the selection process of the training data
- jointly ensure the quality of the data used in the AI system
- control the evolution of the AI's logic and
- investigate whether the AI's conclusion is sufficiently transparent.

If any of these measures are not ensured, then VBE organizations would forgo the partnership and any further investment (Figure 2.4). This is a core principle: "Value-Based Engineering organizations actively consider not investing in a system if there are ethical grounds for such renunciation."

Principle 2: Willingness to Renounce Investment

Figure 2.4: Ethical forgoing of profits.

Value-Based Engineering is about an open and honest stakeholder dialogue

The example of the university admission system brings a third principle to the front that is essential for VBE and IEEE 7000™: the inclusion of stakeholders. How would students perceive a university (or a company), when they learn that their diligently written motivation letters are only read by an AI system? How does it feel to know that your own care and motivation is scrutinized in this way by a non-living void of neutrality? Students in this university case are the direct stakeholders and VBE with IEEE 7000™ recommends asking them. Would students like their motivation letters to be received in this way? What values are borne by such a practice? Neutrality? Efficiency? Absence of care? Perhaps, in contrast, justice? Or, fair unbiased application treatment? The list of potential negative and positive values shows that analyzing students' letters with an AI is an ethical challenge. There are diverging views and hopes. Furthermore, the views of the indirect stakeholders, such as, in this case, the ministry of education, the dean's admission office and the professors teaching the students admitted in this way, also play a role.

Many organizations still shy away from an honest weighing of external stakeholder views, in a critical dialogue. They feel uneasy about dealing with critical voices that may undermine their freedom to make their own choices.[12] However, the intuition to fear critical voices is a clear indication of ethical ambiguity. When innovation teams feel uneasy to openly and honestly discuss their system ideas with critical stakeholders, it should be a warning sign for them that they might have something to hide. VBE with IEEE 7000™ resolves this negative tension. No business in the service of customers should feel uneasy about its practices. Uneasiness is like a snake eating up the motivation of all parties involved in a project. Therefore, the third principle of VBE organizations is to envision and plan their systems in honest and open cooperation with an extended group of direct and indirect stakeholder representatives, including critical ones (Figure 2.5).

Principle 3: Stakeholder Inclusiveness

Figure 2.5: Stakeholder dialogue inspired by ideas and facts.

Value-Based Engineering uses moral philosophies and respects spiritual/religious traditions to elicit values

Understanding the wide spectrum of thoughts and reactions of stakeholders to one's own system idea or early concept of operations is extremely valuable in anticipating all kinds of value breaches as well as positive value potentials. In three case studies, it was found that conventional product roadmapping sees few human and social values impacted (Bednar & Spiekermann, 2021). Some, like privacy and security, have, in recent times, tended to be on system developers' radars more often, but research shows that normally no more than four to seven values are covered in technology roadmaps. In contrast, when people are engaged in value-based thinking, with the help of the ethical frameworks as they are promoted in this book, creativity around potential value impacts increases significantly: each involved stakeholder then identifies between 16 to 19 positive or negative values per technology case.

It was also found that reflecting on values helps with seeing the potentially adverse effects of a system's deployment. Various case studies suggest that innovators building technology roadmaps with a function-driven mindset end up discerning almost no negatives to a project. In contrast, innovators working with the ethical frameworks used in VBE and IEEE 7000^{TM} identify, on average, 10 negative value risks per person involved in the project (Bednar & Spiekermann, 2021). That said, the kind of value elicitation used in VBE and IEEE 7000^{TM} is not just asking for any kind of stakeholder preference. It is not a simple harms/benefit brainstorming exercise. Value elicitation in VBE is guided by established ethical frameworks and by the spiritual/religious traditions stakeholders might have (Figure 2.6). Only by such a sound moral philosophical base, innovators can judge whether their innovations are "doing good" and show care. The fourth VBE principle therefore says: "Value-Based Engineering organizations use moral philosophies for value elicitation."

Principle 4: Use Moral Philosophies for Value Elicitation

Figure 2.6: Understanding ethical system implications through philosophers' perspectives.

The three established ethical frameworks used to elicit values for VBE and in compliance with IEEE 7000™ are Utilitarianism, Virtue Ethics and Duty Ethics (in the same order given here). First, we anticipate the harms and benefits that could result for the direct and indirect stakeholders, if the system were ubiquitously deployed. This is the utilitarian perspective, which provides for a very broad perspective on any consequences the system might have. Second, the virtue effects on human users are questioned. Virtues describe the habitual character quality of a person that makes her or him a good and moral community member and decision-maker. Or, in simpler terms, one could say a virtue is "the positive value of human conduct" (p. 24 in IEEE, 2021a). Examples are humbleness, moderation, kindness, attentiveness, reliability, etc. These human values are often undermined by timely IT systems, and it is a particular concern of VBE that systems should strengthen human virtues rather than undermine them. It is important to anticipate the long-term character effects of an IT system, imagining what would happen if the system were used at scale. And thirdly, the question is asked of whether there are any duty-ethical principles touched upon by the IT system and potentially already seen by utilitarian or virtue-ethical reflection that are of such universal importance that they should be treated with particular care in the system's future design.

Committing to VBE means to commit, at the very least, to these three ethical frameworks of moral philosophy. This is not only because these three ethical frameworks are the most established ones (covered probably in every ethics class in the Western world), but also because our research has shown that these three philosophies are complementary in their ability to unveil values (Bednar & Spiekermann, 2022). The general utilitarianism we use in VBE allows us to zoom out and see a future SOI and its wider societal implications from a bird's-eye perspective. Virtue ethics allows us to specifically unveil culturally founded expectations on human long-term conduct. And, duty ethics allows us to pull out a stakeholder's personal "maxims," which they want to see respected in a system, for higher reasons. Different cultures have different expectations and views on what is good conduct, and they also have different maxims and higher reasons for wanting a system to be a certain way. To ensure respect for local and regional traditions, it is, therefore, important to conduct a virtue-ethical and duty-ethical reflection.

However, many regions of the world still have strong spiritual traditions in which certain values might be cherished that are foreign to a Western style of thinking, and that might not be left out by following utilitarianism, virtue ethics or duty ethics. Therefore, IEEE 7000™ recommends questioning whether a region of system deployment has such a tradition, and if yes, to discuss the long-term value impact of a system against the background of that tradition.

Note that the philosophical frameworks are used to reflect on the "long-term" (5–10 years) value implications of a system that is imagined to be "ubiquitously deployed" or used "at scale." This practice of envisioning a system's effects at scale is also embraced by Value-Sensitive Design (Friedman & Hendry, 2012b). The long-term and ubiquitous perspective is vital. Our Institute's research suggests that imagining a system to be deployed at scale or to be a future monopoly makes people think more carefully about the system's potential value implications, than if this assumption was not made. Many negative value potentials only materialize when a system has a large number of users or a dominant market position.

Value-Based Engineering is context-sensitive instead of promoting value lists

Assuming a dominant market position for a service and analyzing its value impact against this economic background is only one pillar of value elicitation. The second pillar is a bottom-up, unbiased, context-driven and ideally physical exploration of the value space. "Unbiased" means that VBE recommends not using existing value lists or principles for the initial value exploration phase.

In recent years, many such lists have been published by leading institutions around the world, emphasizing values such as justice, privacy, equality, transparency, etc. (Jobin, Ienca, & Vayena, 2019). These principle lists show that organizations around the world have made a commitment to embrace more values and ethics in their IT systems, an important step that should not be underestimated for its effects on corporate decision-making. However, when looking at IT projects in the field where the innovation teams and engineers tried to fit the logic of value lists to specific technology contexts, the inflexible and generic principle-based approach often did not work. Take the values of privacy or wellbeing in the EU's Assessment List for Trustworthy AI (HLEG of the EU Commission, 2020). There have been grand military projects trying to use the ALTAI list to weed out potential privacy and wellbeing issues in their military system; unsurprisingly, they are having a hard time finding any, because in the battle context – for instance, on a fighter jet – privacy and wellbeing are simply not what the pilot is most concerned about. In another retail-related project, the innovation team was so focused on embracing privacy as a core issue that it was blinkered to the true concerns of retail customers, which included tangible help and convenience, rather than privacy. Neither help nor convenience are values included in any of the 84 most well-known value principle lists for AI published in recent years (Jobin et al., 2019).

Value issues are also so prevalent in each technology case that value lists are but the small tip of a project's real iceberg of potential issues and opportunities. In the various VBE case studies conducted at Vienna University of Economics and Business, more than 10 core value clusters were typically identified, each one of them being again composed of multiple instrumental value qualities. In a trial with a Vienna telemedicine platform, for instance, 93 values were directly or indirectly mentioned by stakeholders, forming 13 core value clusters. In another project with UNICEF, 10 value clusters were identified, based on 56 values originally named by stakeholders. All of these values are deeply and contextually bound to the system itself and to the locus where it will be deployed. Only a fraction of this relevant value-spectrum can be found in officiall value lists.

Against the background of this evidence, it seems valid to conclude that the use of value principle lists for VBE is only marginally productive. In the initial value exploration phase, they should not be used (Le Dantec, Poole, & Wyche, 2009) (Spiekermann, 2021a). Instead, it should come as no surprise that the fifth principle of VBE is that innovation teams should strive to deeply understand the context of their systems' deployment and anticipate its effects (Figure 2.7).

Principle 5: Context Sensitivity

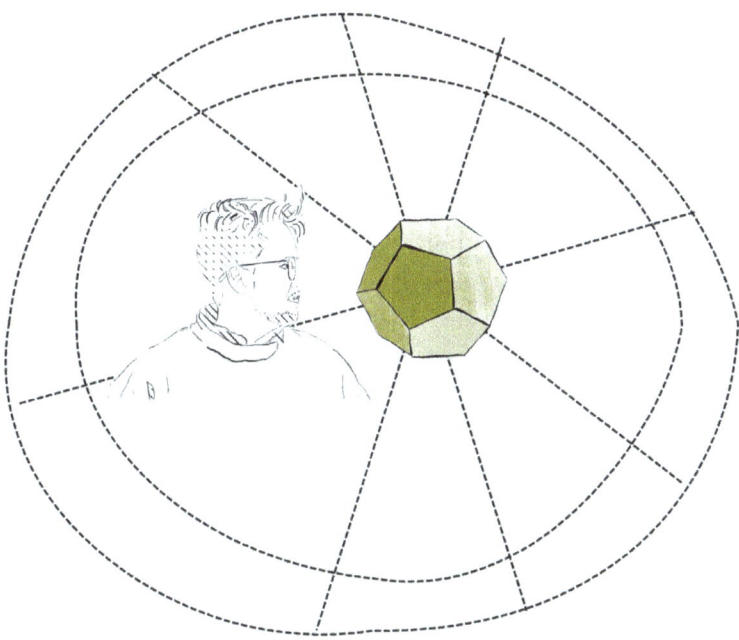

Figure 2.7: The context of a technological thing is key for value analysis.

That said, value principle lists should still be regarded as key hygiene factors to watch out for, when building any kind of system. Consider once more, the EU's ALTAI List. The list says that humans should have control over AI systems (agency). Systems should be safe. Data flows should be protected and secure (privacy and security). Algorithms' skewness towards certain outcomes, such as, for example, discrimination against black people (lack of fairness), should be made transparent and false judgments, of course, avoided. All in all, algorithms should hopefully be in the service of social and environmental wellbeing. One could argue that if such qualities were not in place, then any system (and not just AI) would actually be so suboptimal at its very operational core that it would be hard to operate, sell and maintain in the long-run. Who would be willing to bear responsibility and run the operational risk of a system today, where these qualities are not ensured? Value principle lists do not make a system particularly "ethical" or "valuable"; they make a system good enough to stay in the market at reasonable operational, legal and human cost.

Value-Based Engineering ensures a respect for regulation

The legal and operational costs of a system segue into another important VBE principle: Tech companies wanting to act globally should also be thinking more locally. They should be ready to limit their economies of scale to some extent, in order to cater more seriously to regional interests.

The first 25 years of digitization between the mid 1990s and today saw the rollout of some almost hegemonic "winner-take-all" platforms, hardware and software systems. Only a few global market regions that wanted to maintain their local control over digitization, such as China and Russia, succeeded in building their own regional monopolies or oligopolies. In general, one could say that those who made the Western world's digital market created law with their code. They determined how people would use the Internet and corporate services. Winner-companies created hegemonic defaults of how things are digitally done today in the Western world; defaults, however, that are primarily designed to ensure the winners' own profit margins (Transatlantic Reflection Group, 2021).

As we are still living in a neoliberal capitalist society where anything goes and shareholder value trumps all, there seems to be nothing wrong with this winner-take-all situation. Companies have a right to be successful, goes the capitalist axiom. Any realistic company listed on the stock exchange, today, will think about how it can maximize its economies of scale and will seek homogeneity in its processes, in order to minimize process cost. Therefore, regional laws such as the EU's Data Protection Regulation or other human rights agreements are not particularly welcomed by tech companies seeing that they undermine their business models and imply reinvestments into an infrastructure that has already matured. Reshaping their systems and business models makes them vulnerable vis-à-vis more flexible and younger competitors, who might be able to comply better with laws at lower cost. Hence, only depending on the nature and rigor of potential legal sanctions do tech companies currently rethink and restructure their data processing operations. The dilemma they are in is that *if* social and human-friendly, moral or regional value structures are not considered from the start of a system's conception, then it is very hard and costly to graft them on, later in a system's life cycle. Sometimes legislation can even imply running separate data processing units in a region, just to comply with that region's legal expectations. Large tech companies are, hence, in a cost dilemma. They underestimated the ethical implications of their systems and business models when they first conceived them, and now, they need to decide whether they continue to spend their money on lawyers to fend off customer and NGO complaints (that were never foreseen in the business plan) or whether they truly embrace a new form of better engineering, for example, by pursuing VBE. Both strategies cost money.

The sixth principle of VBE recommends that organizations should proactively and wholeheartedly respect the ethical principles embedded in the spirit of laws and signed agreements (Figure 2.8). They should embrace the fact that many of their target markets are big enough to merit the respecting of regional laws. They should not prioritize their own continent's system values over and above their local customers' expectations. And certainly, they should not prioritize profit over service quality.

Principle 6: Respect for Regional Laws and International Agreements

Figure 2.8: The law must have an intuition for justice and sanctions.

Value-Based Engineering requires top-management engagement and seeks healthy profit

To not prioritize profit over a service's value quality still seems a bold statement to make in times when shareholder value continues to outweigh all else, in practice. As of 2021, Milton Friedman's (1912–2006) famous claim that "the business of business is business" is still valid. The economist was convinced that the sole social responsibility of a company should be to increase profits, as long as one does not break the law.

In line with a wide range of critiques, VBE scholars would not consider Friedman a role-model thinker for the times ahead.[13] As seen throughout history, cultural and economic perspectives are constantly changing. While greed was still sexy in the early 2000s (German: "*Geiz ist geil*"), this has gradually changed over the past 20 years, particularly after the financial market crash in 2008. In the years since, more sustainable and value-based thinking is on the rise, slowly but steadily replacing the old economy of greed. Famous strategists like Michael Porter have been foreseeing that "the purpose of the corporation must be redefined as creating shared value, not just profit per se. This will drive the next wave of innovation and productivity growth in the global economy . . . learning how to create shared value is our best chance to legitimize business again"(Porter & Kramer, 2011) (Figure 2.9).

"The purpose of the corporation must be redefined as creating shared value." (Michal Porter)

Figure 2.9: Michael Porter.

Against the background of this new economic undercurrent, corporate executives who enter leadership positions are now observed more closely. Those who are found to be too greedy or using tricks to maximize profit at the expense of society don't get away with it, any more. They are increasingly put on personal trial for their behavior. Social networks, investigative journalism, whistleblowers and NGOs expose misconduct. A single mistake can lead to a degree of opprobrium, unknown up to the early 2000s. In times of ever flatter organizational hierarchies, top executives cannot, any longer, hide behind a high position and attempt to spin or legitimize false behavior; or, at least, it has become a dangerous strategy to do so. More and more executives are facing legal sentencing and even jail, regardless of their former career or engagement for their company. As a result, corporate leaders are facing the necessity to develop an old aristocratic skill: they need to work on themselves and their personalities to become virtuous leaders. As the Japanese thinker, Ikujiro

Nonaka, wrote: "[Corporate] judgments must be guided by the individual's values and ethics. Without a foundation of values, executives can't decide what is good or bad" (Ikujiro Nonaka & Takeuchi, 2011) (Figure 2.10). He furthermore pointed out that "[i]n conventional economics, the ultimate goal of any company is to maximize profit. But in the knowledge society, a corporate vision has to transcend such an objective and be based on an absolute value that goes beyond financial matrices."

"[Corporate] judgments must be guided by the individual's values and ethics." (Ikujiro Nonaka)

Figure 2.10: Ikujiro Nonaka.

VBE helps executives and corporate leaders to understand what that value could be, which helps them to transcend profit-oriented thinking. The method, in line with the IEEE 7000™ standard, gives them guidance on how to prioritize the many values that stakeholders mention in response to the operational concept of a new product. They are guided towards considering their own value maxims – what they deem personally to be of universal importance from an ethical perspective. Against this background, the seventh principle of VBE reads as follows: "Corporate leaders engage in introspection and support only those core values as future system principles that they would want to become universal and are therefore willing to publicly and personally endorse" (Figure 2.11).

Principle 7: Leadership Engagement

Figure 2.11: Top managers should be leaders turning the wheel.

There is, however, one important boundary in this argumentation for value-driven leadership: VBE is not against making profit per se. To use the words of the former CEO of German Rail, Heinz Dürr, it is about making *a healthy* profit. A healthy profit is one signaling that a company is in a sustainable state, able to successfully maintain its business mission while paying reasonable wages to its employees and respecting the interests of communities and stakeholders impacted by the business.

Value-Based Engineering operates transparently

Once corporate leaders and their innovation teams commit to healthy profits in the service of human and social value, they will have no problem with making this mission public. They will not hesitate to demonstrate their thinking and arguments; indeed, they will be eager to share it with their employees and the world to motivate the people who work for them and the customers who buy from them. VBE, therefore, encourages innovators to publish an Ethical Policy Statement, which summarizes the core values that an enterprise prioritizes for a product or service.

Such an Ethical Policy Statement is not to be confounded with a marketing slogan or a list of Corporate Social Responsibility commitments. It was explained above that value-based innovation teams run stakeholder groups through an ethical elicitation process, which is focused on a concrete product or service. VBE is not a remote strategic exercise, such as many CSR activities are, today, bemoaned to be. It is also not about a marketing message that is bolted on to a product by a PR agency, after the product has been built. Instead, VBE focuses on the concrete concept(s) of operation for tangible products and services in the early stages of their making. Ethical and value-based thinking is not a general corporate view or a PR promise but rather concrete thinking that enters the product roadmap, the agile development sprints or the developers' list of system goals, very early on.

Ethical Policy Statements can, then, summarize the higher and intrinsic core values that have inspired the system development goals. And, it is the path from these core value principles down to the system requirement practice that should be documented in a separate repository that the IEEE 7000^{TM} standard calls a "Value Register." These two artifacts of value history, the Ethical Policy Statement and the Value Register, help companies to structure, remember and share what they work for. Against this background, the eighth principle of VBE unfolds, which is about the transparency of the value mission, as endorsed in these two documents (Figure 2.12).

Principle 8: Transparency of the Value Mission

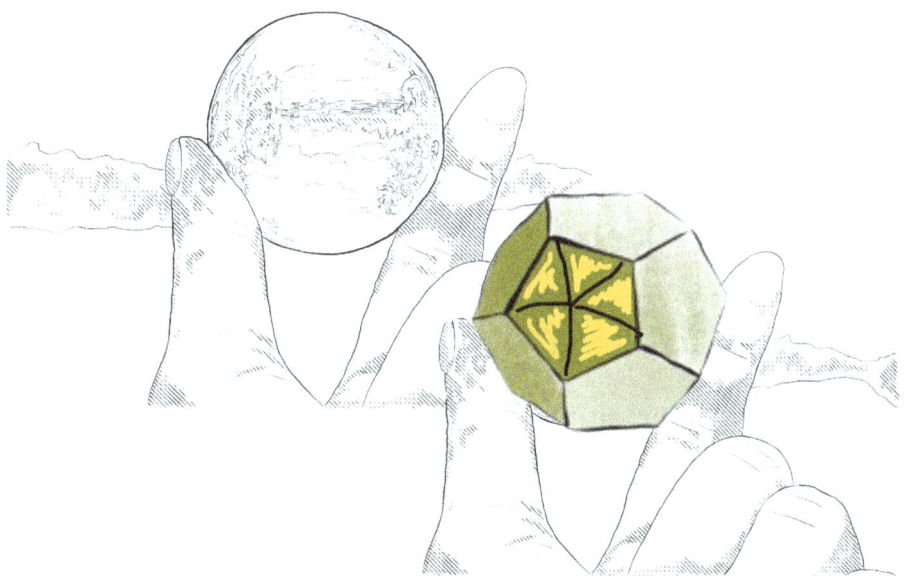

Figure 2.12: IT innovations' value mission is transparently shared.

Value-Based Engineering offers a path to value creation

The Ethical Value Register contains a path from higher intrinsic core values (principles) to system requirements (practice). It is the transparency tool that is also standardized for information management in IEEE 7000™ (IEEE, 2021a). But what does this path to value creation look like, and how is it pursued concretely?

Most companies that build technology today and operate with professional maturity are following processes or work flows that they have either defined for themselves or adopted from industry standards. VBE is therefore equally process-driven. It has been one goal of the IEEE 7000™ standardization effort to understand how its processes can be aligned with activities that are prescribed by widely used process frameworks, such as ISO 15288 (ISO, 2015) or other established system development life cycle models (so called "SDLCs"). (ISO/IEC/IEEE, 2017; Spiekermann, 2016). Established SDLC frameworks outline how systems should be engineered, step-by-step. They define what goes into processes as inputs and what comes out of them as outputs and/or outcomes. They show what actors are involved in what roles and what activities and tasks are completed. VBE is not only an equally detailed and, thereby, reliable and repeatable method, but it is also designed to be accommodative with such established corporate practices.

That said, VBE still has its own orderly blocks of work, activities and tasks. In fact, the path to value creation can be summarized with the help of three grand process blocks: a block of concept and context exploration work, a block for the ethical exploration and value prioritization stages and a third block of activities, where the ethically aligned design of an SOI is created (see Figure 2.13). In this latter part of work, core values of a system are translated into practical system of interest (SOI) requirements.[14] These process blocks and processes can be run through in an iterative, repetitive and interlinked way.

Concept and context exploration

Before a VBE project can start, a number of activities and tasks need to be completed, in preparation. Most importantly, an understanding of the SOI's initial set-up, its context and the likely challenges associated with legal, social and environmental feasibility must be gained. An organization will need to graphically depict the components of the SOI in a concept of operation; for instance, by using box diagrams, contextual diagrams, high-level UML sequence diagrams, etc. Stakeholders are identified and studied in terms of their expectations of the system. Data flows and ethically relevant system boundaries as well as the control the organization has over its envisioned external partners are analyzed. Given these tasks, one could draw an analogy to garden planning, where it is envisioned how flowers and trees (components) are to be brought

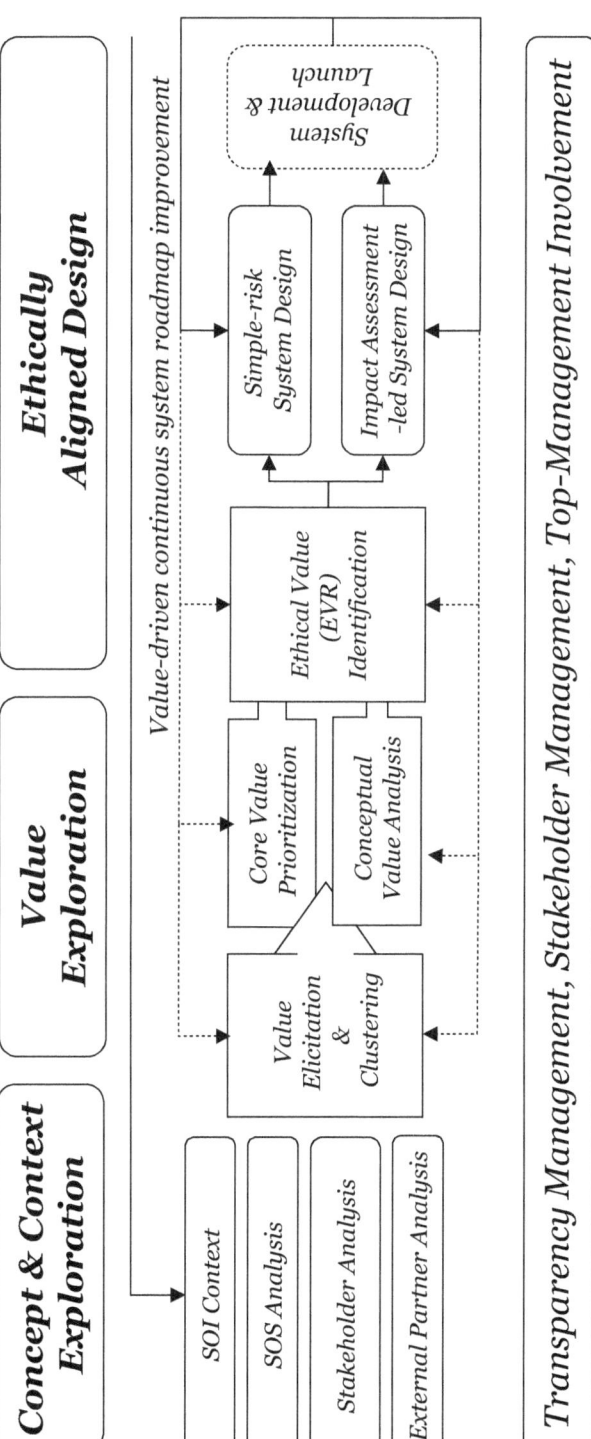

Figure 2.13: Value-Based Engineering (VBE) process overview.

into the soil, how water (data) flows, how people are going to use and maintain the garden, etc. This is why Figure 4.2 introduces the depiction of a garden, which resurfaces throughout this book. VBE invites innovation teams to liken themselves to gardeners. And, one of the first messages of VBE is that you cannot just plant any system into any context, just as you cannot grow any plant in any climate (see Figure 4.2).

Ethical value exploration

Once such preparatory analyses are completed and an innovation team with stakeholder representatives has been appointed, the organization building the SOI can start with the ethical exploration. Through value exploration, positive and negative values relevant for an SOI are identified and prioritized. The deployment of the system needs to be envisioned at scale. The harms and benefits that could result for a broad set of direct and indirect stakeholders if the system were ubiquitously deployed (utilitarian perspective) are anticipated, during this phase. The virtue effects on human users are questioned (virtue-ethical perspective), and the question is asked of whether there are any duty-ethical principles touched upon by the system that should be treated with care in the system's future design (duty-ethical perspective). Together with a potentially regional ethical framework, these ethical perspectives support the identification of the core value clusters that need fostering or protection in the SOI.

Organizations then engage their top management to prioritize these core values, each of which is conceptually built up by value qualities relevant to the SOI.[15] Prioritization is not profit-driven. Instead, core value priorities are informed by the comparison of alternative core value missions that may be supported by the SOI. Prioritization is supported by existing ethical frameworks, such as human rights agreements, existing regional legislations and corporate social responsibility (CSR) commitments already made. Additionally, value principles lists, such as those critically discussed above, can play a part here. Value priorities should mirror the respect the organization has for people, society and the environment. Top management should play a vital role in this value prioritization activity.

A vital part of the ethical exploration is to understand the core values prioritized in depth. Stakeholders in their bottom-up and context-sensitive dialogues are, more or less, able to express in what respect they find certain values important. But they usually do not have the knowledge or bird's-eye perspective to really understand the conceptual details of a value. Take the example of privacy. Stakeholders may state that they are concerned about the security of their data and want to control the further use of their personal data for secondary purposes. A VBE project will need to respect such concerns. But if the project goes ahead and prioritizes privacy as a system's core value, then it will also need to go beyond what stakeholders saw and said. They will need to query what privacy experts and the legal world would want to see in a system that is later marketed as being particularly privacy-friendly. Through the conceptual value analysis (or call it "expert-view") of a value, additional value qualities that a normal stakeholder or project team member would not have seen or mentioned come into play. In the privacy case, for instance, legal issues such as data portability, privacy by design, etc. might be identified as relevant.[16] For this reason, VBE teams engage in a conceptual analysis of their prioritized core values. As Principle 9 states, they need to understand their core values in depth. This is illustrated in Figure 2.14, which shows how a value (tetrahedron) conceptually consists of multiple quality sides, which are again linked to multiple value dispositions (indicated by dotted boundaries).

Principle 9: Understanding Values in Depth

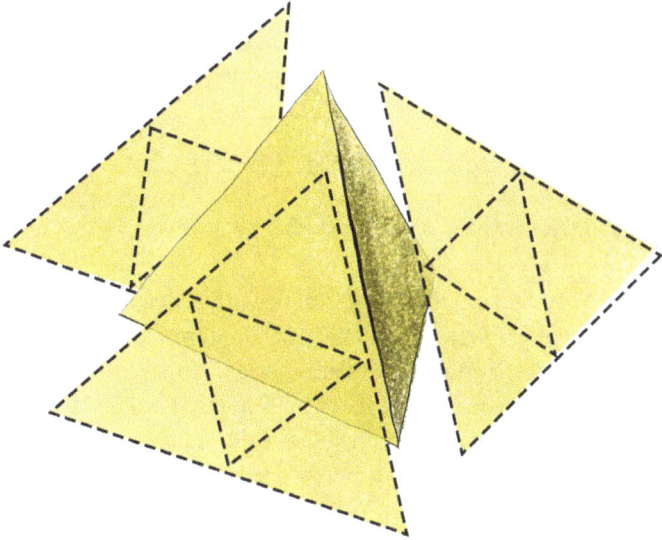

Figure 2.14: A tetrahedron symbolizes a value with value quality faces.

Ethically aligned design

Once core values are prioritized and analyzed, their alignment with system requirements can start. All sets of prioritized core values with value qualities are first translated into so-called "Ethical Value Requirements (EVRs)," which are qualitative descriptions of what should be done to actualize a value quality. These EVRs are then translated into system requirements. Principle 10 of Value-Based Engineering requires organizations to perform this translation with the help of a risk rationale (Figure 2.15).

Principle 10: Using Risk-Analysis for System Requirements Elicitation

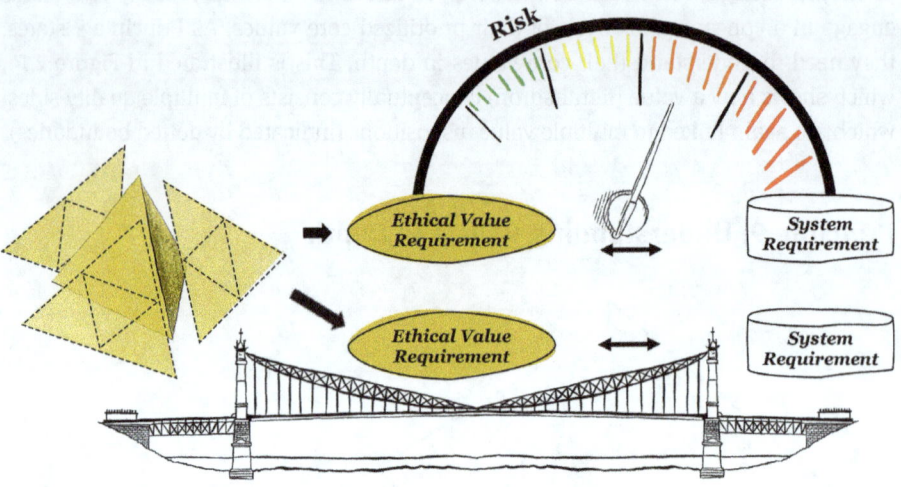

Figure 2.15: A risk-based logic translates EVRs into system requirements.

Normally, EVR translation into system requirements is done with a simple risk-based approach; that is, with the help of a threat-control analysis. The threat-control analysis scrutinizes each EVR to see how it could be undermined or "attacked." In other words, it is being asked whether there is a threat that the EVR might not be reached. Then, in a second step, how a potential undermining of an EVR could be mitigated is considered. Mitigation implies either concrete technical measures taken or system policies in place. The expectations of partners are clarified as well as partners' operational restrictions and service levels. In short: One or more organizational and technical system requirements are identified for each EVR. The technical system requirements are then integrated into the organization's existing functional product road map, which is managed by the technical development units that are tasked with bringing the system up and running.

Risk assessment-based design

Sometimes core values and their value qualities may be of particularly high importance for an organization investing in a system. For example, privacy and security are values of the highest importance for banking systems, whereas reliability and transparency may be particularly important values for financial trading systems. Likewise, ensuring patient health and safety is crucial for any medical or body-attached system. VBE (unlike IEEE 7000™) recommends that organizations analyze where particularly high value-expectations or liabilities are associated with an SOI. And, if there are such "high-risk" values, then the organization should embrace not only a simple threat-control-based risk design, but engage in a proper impact assessment-based system design. Impact assessment-based design is also a "risk-based" system design approach, but it is much more rigorous than a simple threat-control analysis (see, for example, the NIST standard for system security (NIST, 2013)).

Ethically aligned design needs iterations and adjustment

Finally, it is well known that new products and services are regularly put to unexpected uses. In most cases, once an SOI is deployed, it is not used in full alignment with the intentions of the engineers. Negative as well as positive value effects appear where nobody expected them. Think of the Facebook "like" button as an example. It was introduced by the social network with the good intention of giving users the possibility to provide their peers with positive feedback; only in hindsight, did it become clear that this like button can breed addiction to the service – as well as an unexpected amount of envy on the platform (Krasnova, Widjaja, Buxmann, Wenninger, & Benbasat, 2015). When a system causes such unexpected value harms, then a VBE process foresees iteration. A value, such as the avoidance of envy, is inserted as a new system priority. This priority is then conceptually analyzed and EVRs are identified, triggering system requirements that mitigate the unexpected value harm. Organizations should therefore constantly monitor the market's value reactions to the system and then be ready to adjust their system design, accordingly.[17]

Check questions

- What are the drawbacks and benefits of value-principle lists?
- What does VBE do to accommodate different cultures?
- What is the role of stakeholders in VBE projects?
- Why can companies not conceive of VBE as "bowling alone"?
- Why can the garden metaphor be useful in relation to IT innovation?

Chapter 3
What values are

The term value is derived from the Latin word "*valere*," which means "to be strong" or "to be worthy." Values seem to be at the core of cultures and of the innovations brought forth by them. They allow people and organizations to frame the meaning and ends of their actions. However, despite their central role in human behavior, what values *are*, precisely, is not common knowledge.

In our modern economic world, the value of something is often equated with money. But this equating of value with money overlooks the fact that money is only a tool facilitating the exchange of products and services that are the actual bearers of value. So what are values? And, how can they be instantiated in a computer system?

Towards a definition of values

Defining values is not an easy endeavor in our modern times, because values are not tangible things and, therefore, cannot be physically measured, inspected or touched. Instead, they force our scientific mindset to venture into an invisible phenomenon of symbolic meaning.

The biggest stumbling block in understanding and working with values is the idea that they are individual preferences primarily originating within peoples' thoughts. Modern definitions of value(s), such as the one in the Oxford English Dictionary, encourage this view. The dictionary defines values as "principles or standards of a person or society, the personal or societal judgements of what is valuable and important in life" (taken from p. 23 of Friedmann, 2019). Other contemporary philosophical definitions of value(s) are along the same line, such as "lasting convictions or matters that people feel should be strived for in general and not just for themselves to be able to lead a good life or realize a good society" (p. 1 in van de Poel, 2018).

Such subjectivist definitions of values as personal judgments, convictions or, perhaps, just personal opinions does not hold true, upon deeper scrutiny, though. Take the value of privacy as an example. Many people might be convinced that privacy has ceased to be important in a time when ubiquitous computing and social networks see more data exchange than ever before, in human history. Some societies, like China today, might even establish a norm of citizen surveillance, thereby seeming to refute the value of privacy. But such personal preferences or societal norms do not make the value of privacy disappear. The fact that the value of privacy continues to be discussed – no matter anyone's convictions or social norms – is an indicator that values, as such, are somehow independent phenomena that exist a priori of any individual's or group's judgment. As Max Scheler famously put it in his magnum opus on Material Value Ethics (p. 77), the world and its value principles

"The ego is neither the point of departure...nor the producer of essences [like values]" (Max Scheler)

Figure 3.1: Max Scheler (1874–1928).

do not depend as much on human thought and opinions as we often like to believe: "... the ego is neither the point of departure for the apprehension nor the producer of essences," he wrote (Scheler, 1921 (1973)) (Figure 3.1). He saw values as "the ultimate stuff of our moral consciousness; they are the material to which moral consciousness *is directed*; they are the intentional *objects of acts* of feeling, or conscience, or moral consciousness" (p. 7 in Kelly, 2011).

Embracing Scheler's objectivist understanding of values, Nicolai Hartmann described how we might, therefore, think of values in similar ways as we think of geometric principles existing in our cosmos. We know objectively what an ideal triangle is, in principle, and can describe it with the Pythagorean Theorem. And, when we see a thing with a triangular shape in reality, we recognize the objectively given principle of the triangle in it. This is not a matter of personal conviction or judgment. The same goes for values. We see that the Gestalt of something is good or right. For example: courage. When we observe someone being courageous, we immediately recognize the principle of courage in the action, even though courageous actions can be very diverse. Hartmann defined values as "principles of the ought-to-be" (Figure 3.2), which we intuitively perceive and to which we have given words in our languages.

That said, in perceiving an objectively given value like courage or privacy, it is still true that individuals bring in their own subjective history, culture, character and knowledge. People differ in how they individually recognize positive values as something desirable and equally diverge in noticing the absence thereof, such as a disappointing lack of courage in someone's behavior or a lack of privacy in a surveillance situation. And, as a consequence, people react differently. Many will applaud the courageous or avoid the privacy-intrusive environment. Their behavior

Chapter 3 What values are

Values are "principles of the ought-to-be"
(Nicolai Hartmann)

Figure 3.2: Nicolai Hartmann (1882–1950).

responds to the existence or absence of values. Humans, values and the things that bear them are ontologically inseparable from the start.[18]

In line with this conception of values, Harvard anthropologist, Clyde Kluckhohn, defined values as "conceptions of the desirable that influence the selection from available modes, means and ends of action" (Kluckhohn, 1962) (Figure 3.3). Unlike Hartmann, Kluckhohn's definition accommodated for the observation that values trigger emotional responsivity towards or away from the value-bearer. In his definition, though, the word "conception" could be misread to accommodate the kind of subjective constructivism of values refuted above.

In order to align the objective phenomenal nature of values with their individually distinct perception, VBE defines values as follows:

> **Values are phenomena disclosing the degree of desirability of something or someone, giving meaning to and motivating the selection of available modes, means and ends of action.**

By using the verb "to disclose" the *mediating and response dimension* of value perception is embraced (Verbeek, 2016): Values disclose something about the object (person, symbol, relationship, etc.) that bears them. And, they are actualized only through the perception of an observer who apprehends that which is disclosed. The above definition thereby embraces the recent scientific argument that values could also be likened to affordances (Affordance Account of Value Embedding) (Klenk, 2021).[19]

When observers of a value-laden situation are inattentive, however – inexperienced, distracted or strangers to a respective value milieu – they can easily forgo the potential in front of their eyes. They may miss the beauty or reliability of something, for example. In such cases, value dispositions may very well be embedded in a person or a

"[Values are] conceptions of the desirable." (Clyde Kluckhohn)

Figure 3.3: Clyde Kluckhohn (1905–1960).

thing in front of an observer, but they remain dormant as potentials and are not actualized in perception. In the design of computer systems, this happens very often. For example, when functionality for the benefit of a user is buried deep down in a technical menu structure – it is there as a potential, for instance, to improve privacy protection, but nobody uses it. And, therefore, a value like privacy does not effectively materialize.

Why is a rigorous value definition important?

The terminological details of the value definition might seem like over-egging the pudding. Yet, an ontologically misguided understanding of values can create confusion and harm. In fact, misunderstanding what values are can lead to poor policy making, poor business models as well as suboptimal expectations of what makes a system valuable.

Policy implications of a weak value definition

Starting with policy making: In the EU Commission's proposed Artificial Intelligence Act, the term "union values" is referred to, over a dozen times (EU Commission,

2021a). It may be argued that union values should be built "into" systems, so that European technology then "has" these in their design somewhere. But building values physically "into" a system seems physically impossible. How could anyone build an intangible principle tangibly into a product?

Some ethicists of technology suggest that this challenge is just a matter of value definition, and embrace what is called an "Intentional History Account of Value Embedding" (short "IHAVE") (Klenk, 2021; van de Poel & Kroes, 2014). Ignoring the phenomenologically independent and objective nature of values, the Cartesian IHAVE definition of a value is that a system simply has it from the moment it was intentionally designed by an engineer to have it. For example, when an engineer really wants a system to respect user privacy and builds it with a conducive privacy-by-design approach users can potentially benefit from, then IHAVE scholars would argue that the system "embodies" the value of privacy. The personal judgement or intention of the designer is what seems to count for the value *to exist*. And indeed, having good intentions and translating them into a system design is hugely important for the propensity of a system to later bear the intended value(s).

However, one must admit that there should be a difference between a good intention to do something on one side and the ability and effectiveness in doing it on the other. Real value only materializes when it effectively unfolds in the world. And, this is what is ultimately important for companies. Customers cannot live only by good designer intentions and conducive designs. What happens if these are not playing out in the real experience of a product? Too often, intended values simply do not unfold as expected, not even when customers properly follow use manuals. The history of product use shows that intended value is often ignored. Sometimes, unexpected positive value(s) that nobody expected are suddenly appreciated. At other times, negative values that nobody anticipated appear. Since such dynamics are the norm when new products and services are introduced, it seems questionable to credit the ontological existence of values mainly to designer intentions. If this is done, then two factors that are vital for successful VBE are ignored: (1) the effectiveness by which a designer's intentions are actually embedded in a system through respective value dispositions and (2) the role of the system user necessary to perceive the value intended. For these reasons, policy makers seem well advised to insist on proofs of value creation and value protection over time, and not just well-documented compliance proofs that show designer intentions. Value realization should be observed in practice. Value recognition by ordinary users should be investigated. And, value monitoring over time as well as processes to adjust systems with negative value implications should be in place, at least for risky systems.

Business implications of a misguided value definition

A limited value definition is not only problematic for policy makers but also for business. This becomes particularly apparent when business scholars equate values with product features – product features they may even declare to be a "value proposition." The highly regarded strategy consultant Alexander Osterwalder (Osterwalder & Pigneur, 2010), for example, described, in a public lecture, the value proposition of a Tesla car as follows:
- the car!
- a powerful battery,
- a lot of space in the car,
- luxury image,
- fuel for free,
- upgrades,
- great design,
- power range (to drive with the battery) and
- safety.

Looking at this list critically, there is only one true value here and that is safety. All the other points he mentions are mostly prerequisites or dispositions that must be in place in order to create value qualities and values. For example, a powerful battery – if excellently engineered! – can support value qualities such as flexibility, time savings and range, which, in turn, are prerequisites for that value for which a car is ultimately bought: mobility! (Figure 3.4). Upgrades are usually required to increase the reliability of the car and this, in turn, caters to safety. A lot of space is one disposition catering to comfort, but much more is needed to really make a car comfortable. In other words, anyone who esteems mere features or a few components like a battery or upgrades to be "values" has not yet understood the ontological status of true values. She or he risks confusing the technical preconditions of a value with values themselves and, thereby, dramatically underestimates what is actually required by companies to create value. In Heidegger's terminology, one risks prematurely raising "stuff" (German: "Zeug") to be "goods," thereby misleading customers and executives alike.

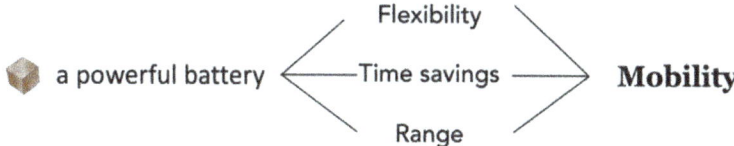

Figure 3.4: Discerning system features from values.

That said, material dispositions catering to values in a system are, of course, important. They are the prerequisites for value creation. Yet, exactly what role they play both for the corporate value proposition as well as for engineering can only be understood after we deepen our understanding of the value phenomenon.

The process of valuation

In our colloquial talk about values, it is often said that we "have" or "hold" certain values. However, strictly speaking, we cannot "have" or "hold" values, since we cannot strictly own an invisible, intangible, metaphysical entity as we have or hold physical things. So what would be a precise way of verbally referring to valuation values?

According to the value phenomenologist, Max Scheler, a person can perceive values in something or someone and can resonate with these. She or he can be attracted to or repelled from positive or negative values "borne by" or "carried by" objects, people, symbols or activities encountered in an environment (Scheler, 1921 (1973)) (Figure 3.6). However, in order for this "valueception" to happen, two things need to be in place: First, the objects and people in the environment really need to be equipped with the right "value dispositions" (p. 79 in Scheler, 1921 (1973)).

Figure 3.5: An SOI bears values.

And second, the value-perceiver must have the value knowledge and milieu experience to actually recognize and correctly classify values. Figure 3.5 illustrates this.

A value disposition is the capacity, characteristic or property "in" an object (or person) that physically provides the potential to enable or inhibit one or more values.[20] Value dispositions are what auditors can, indeed, tangibly find in a computer system. An auditor inspects a computer system and finds for instance that the data stored on a hard drive is or is not encrypted. In the software, she or he will see whether encryption is symmetric or asymmetric, what kind of encryption algorithm is used and what key length. And, seeing these dispositional facts, the auditor can make a value judgment on whether the system is secure or not.

Note that this value judgment is not subjective or a personal preference. Whether an algorithm can be considered secure or not depends on best available techniques (BAT) and standard recommendations for encryption at a specific time and in a specific context. An auditor will need to know about these. He or she will need security (value) knowledge, in addition to some (security) milieu experience, in order to judge a system as either secure or not, based on factual properties (Figure 3.6).

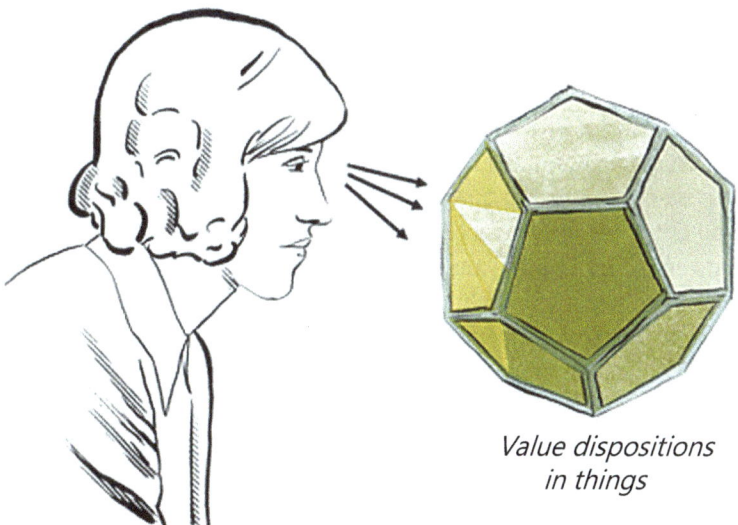

Value dispositions in things

Figure 3.6: Perceiving the value of a thing by judging on value dispositions.

Technical valuation is not a matter of personal taste

Let's give the auditor the name, Annie. Annie has privately been part of a hacker community since her teenage years and has inhaled a milieu of extremely high-level "security speak." So Annie's long-term "life-form" (in Wittgenstein's sense (Wittgenstein, 1993)) has led her to develop a relatively elaborate knowledge of security. As she audits a bank's encryption system, she may be relatively tougher in her judgment on the system's security dispositions than an alternative auditor, Paul, who has just passed his first security-officer certificate, but who has not spent his life in the domain. Annie might, for instance, find the encryption key not strong enough. She might disagree with Paul, who has learned that the law accepts 128-bit key length as adequate for meeting the legal security standard. Despite this guideline, Annie thinks it is not secure enough for bank customers, and therefore recommends the bank uses a 256-bit key length. This judgment of the right security level is not her "subjective" judgment, even though she expresses it as a "subject." Instead, her judgment on how to ensure the bank's security would probably be shared by her peers from the hacker community. It is "intersubjectively" true.[21] That said, the inexperienced Paul does not share Annie's value judgment. What the example shows is that a given value disposition in a system (the key length, the algorithm's symmetry, etc.) can obviously be valued in different ways. So, are we back to subjectivity and preferences, then? Should Annie have her opinion and Paul his and both are equally right and respectable? Certainly not!

Annie and Paul differ in how they evaluate the relationship between the (banking) system's factual dispositions and the security quality potential they entail. This relationship can be analyzed quite objectively (Figure 3.7): First, the potential for the system to be considered secure is objectively tied back to tangible system dispositions: A longer key is harder to decipher than a shorter one and, therefore, has an indisputably higher quality potential for security. An asymmetric encryption system is harder to attack than a symmetric one and, therefore, has an objectively higher security quality potential, as well. Hence, the material properties of the thing (the value dispositions) enable or hinder value quality potentials to a different degree. Second, context factors have a relatively stable influence on value potentials. The context of banking, for instance, is specifically vulnerable when it comes to protecting customer security. The social constituency of the banking context, therefore, means that the materialization of security requires particularly sophisticated protection properties (elaborate tangible value dispositions) to resist attacks.

An expert like Annie simultaneously integrates these two value dimensions in her judgment (Figure 3.7): First, the knowledge on whether system dispositions are sufficient to actualize a value and second, the knowledge of the contextually co-determined value unfolding. It is, hence, not her personal convictions that dominate the security judgment, but objectively given instances in the world. And, it is likely that if Paul had the same experience knowledge (or milieu) that Annie has, he would probably reach the same conclusions as she does. So, with experience, both are likely to converge towards a shared value truth.

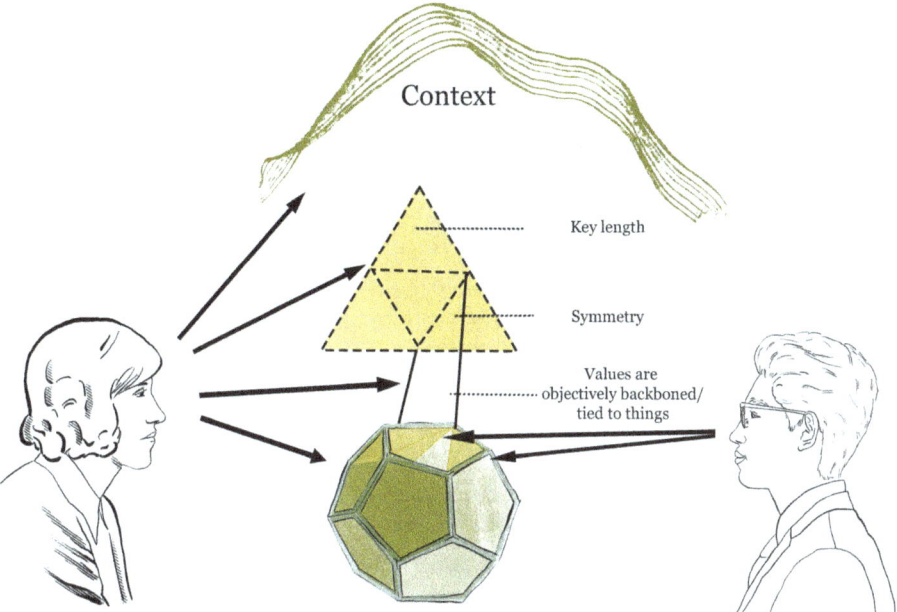

Figure 3.7: Experts and laymen judging value.

The importance of experience in value quality judgments

Is Paul an inferior auditor? What is it that can make his value quality judgment poorer than Annie's?

Annie is able to simultaneously integrate the contextual knowledge relevant for the value judgment, including the technical dispositions that the value of security is tied back to. It is not that she mentally weighs each one of these factually given dispositions (such as encryption strength) and then calculates some kind of security value score. Even though some representationists might theoretically describe the cognitive exercise of valuation in this way, this is not what can be confirmed to happen (Hobbs, 2017) phenomenologically. Long-term research in attention psychology has shown that only beginners in a field (like Paul) need to consciously call up their declarative knowledge of something (like the crypto dispositions here) and then cognitively weigh the facts they observe to reach a conclusion or initiate an action. In contrast, experts (like Annie) who have learned a skill have procedural knowledge to act in a situation (Taatgen & Lee, 2003); this is what Polanyi termed "tacit knowledge" (Polanyi, 1974). Phenomenologically, they have the skilled intuition to recognize one or several value qualities of a thing as a whole (Figure 3.7) and have learned to tend to them, similar to the way in which an experienced car driver uses various gears automatically, when accelerating a car.

So, does the holistic perception of value qualities that experts like Annie possess make them better assessors of values? Yes, indeed. And, there are at least two reasons for this: First, inexperienced beginners run through the sequential stages of declarative knowledge application, and in doing so, their performance is typically slower and more error-prone than that of skillful milieu experts (Taatgen & Lee, 2003). In his accumulation of relevant security value dispositions, for instance, the inexperienced Paul is likely to make mistakes: not attending to some or forgetting about other issues, and combining them in a suboptimal way or falsely judging their relevance for ensuring security. This can degrade his assessment of the given value. Secondly, Annie intuitively observes the security disposition and value qualities as "wholes," or what psychologists call "Gestalten" (von Ehrenfels, 1890). Annie acts vis-a-vis these mentally integrated value wholes, and this is not only more efficient as observed in attention research – Gestalt psychology also recognizes how wholes are typically greater than their parts. In a Gestalt, it may, for instance, be the constellation of parts that makes a quality difference. And, as Annie's expert knowledge allows her to recognize the Gestalt of the bank's security, she might, therefore, recognize issues that Paul could not even recognize if he wanted to.[22]

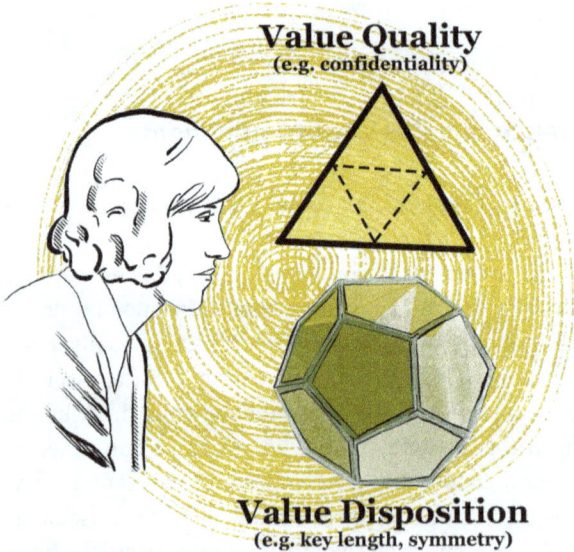

Figure 3.8: Two ontological value layers.

Against this background, we can distinguish two ontological value layers for a thing like a computer system: a lower material layer of value dispositions, such as key length or symmetry that can be touched and inspected, one by one; and a higher value layer of qualities, which are objectively tied back to these dispositions and which are recognized by milieu experts as wholes (just like Annie is judging the computer confidentiality as a whole).[23] This is illustrated in Figure 3.8.

Towards a three-layered value ontology

In thinking about Annie and Paul, we have seen that the two are responsive in a more or less sophisticated way to what are called "value qualities" – value qualities that are, to some extent, objectively tied back to concrete value dispositions in a thing. What a security expert might have noticed, though, is that so far, the encryption example only referred to one specific value quality relevant for security, which is confidentiality. IT system security, however, is not achievable through confidentiality alone. System security is typically created only if other value qualities are present as well, such as content integrity and service availability. It must ensure that no viruses can corrupt customers' account information (example for integrity), and it needs to make sure that customer accounts are always available. For security audits, a recognized trio of value qualities that constitute system security are referred to as the "CIA" principles (ISO, 2014). Thus, it is actually multiple value qualities taken together that constitute one value (see Figure 3.9).

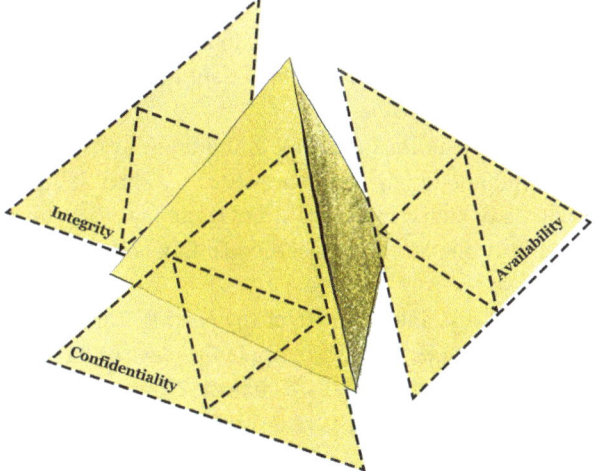

Figure 3.9: Multiple value qualities constitute the core value of security.

As Annie screens the system, the varying value qualities come to her mind. When she sees that encryption is weak, for example, and that the bank, thus, offers poor confidentiality, this observation can easily influence her expectations of other value qualities like integrity. If it is so easy for an attacker to decrypt customer data, then what would stop him from altering a deciphered file, corrupting its integrity, re-encrypting it and leaving the system looking normal? Edmund Husserl, the scientific father of phenomenology, described how we often have a mental awareness of what is at the back of a thing we initially only see the two-dimensional surface of. Due to our knowledge and experience, we already anticipate what is next or what is behind. Husserl described this process of subsequent detection of reality as "progressive self-giving" (p. 115 in (Hobbs, 2017).

It is for this reason that a tetrahedron is used in this book to depict values (Figure 3.9). A value (like security) has multiple sides or qualities to it that give themselves progressively to observers. This is similar to holding a tetrahedron. Even though observers only see one or two faces at a time, the entire solidness is, nevertheless, apprehended. This solidness of the platonic tetrahedron is similar to the wholeness which we speak of, when we refer to values. Very typically we would talk, for instance, about the security of a system as a whole, even though, precisely speaking, we only thought of one value quality, such as confidentiality. It is within this wholeness of perception that value qualities give themselves progressively to consciousness.

The example shows that there is a difference between the overall value of security and various distinguishable value qualities that progressively emerge in our experience and awareness of things (confidentiality, integrity, availability, etc.). The overall value reveals itself in one "unitary experience" of various progressively unfolded value qualities (p. 104 in Hobbs, 2017).The overall values, which are called "ideal core values" in VBE, are often the ones we generically refer to when we speak about value phenomena. We call for security, privacy or freedom of systems for instance. We generically refer to union values, or values so important that we consider them a human right.

Taking ideal core values and value qualities together and adding to these the physical value dispositions described above (e.g., the encryption), a three-layered value ontology becomes apparent, as depicted in Figure 3.10[24] (Note that this value ontology is of philosophical nature. It needs to be discerned from what computer scientists call "ontology"[25]).

Ideal core values are important lead principles for system engineering. Take values beyond security, such as beauty, knowledge or friendship. Apple Inc. devices have set a global benchmark for the beauty of computer systems, which made it one of the most valuable companies on earth today. Knowledge is a core value sought by an encyclopedia like Wikipedia. And, social networks continue to fuel heated discussions, because, on the one hand, they are used for friendship and on the other hand, they undermine that same sense of community through hate, envy and miscommunication. Perhaps it would have been good for a company like Facebook to reflect a bit more on the ideal value of "friendship" before just "connecting people."

One philosopher who significantly contributed to our understanding of ideal values was Nicolai Hartmann (Figure 3.3). In his work on ethics, he detailed the nature of many of them (Hartmann, 1932), such as goodness, nobility, abundance, purity, justice, wisdom, courage, prudence, charity, honesty, loyalty, trust, humbleness, etc. Hartmann

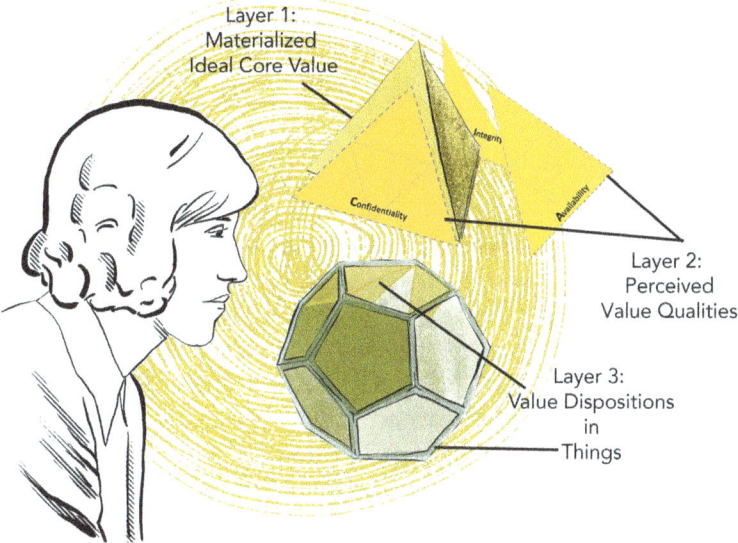

Figure 3.10: Three-layered value ontology.

was a Platonist. He argued that values are ideal principles that have a form of being that is ontologically objective. But this objective ideal is not entirely perceptible to human observers. Ideal values in Hartmann's understanding are just as Plato's eternal ideas in the Cave analogy. We humans, according to this Cave analogy, are only able to observe shadows of the eternal ideas from the perspective of our cave. Or, in other words, while we grasp the core idea about a value what we actually see is really only a context specific selection of its qualities. One real-world example that illustrates this is the value of beauty.

Ideal values materialize as complex value quality structures

Imagine a live performance of Beethoven's Sonata op. 2 No. 3; a sonata that has been recognized for its musical beauty. While the sonata may be theoretically recognized for its ideal beauty, its real beauty really depends on a multitude of value quality factors that need to be in place for this beauty to materialize. First, there is the pianist (value bearer 1) whose clarity and precision of play as well as correctness and gentleness of touch are vital interactive qualities for the listener to actually enjoy the beauty of the piece. Furthermore, the piano (value bearer 2) needs to be correctly tuned. Professional pianists often use a grand piano, because the volume and shape of this bigger instrument (its value dispositions) mean that the beauty of the piece can come about even more, enabling a range and fullness of sound that is not achievable with an ordinary piano. Finally, the pianist's and piano's dispositions come together with the structure, the pitch and the sequence of tones in the score (value bearer 3). These value dispositions of the piece itself are responsible for value qualities like the perceived harmony and transport in listening.

This example makes plain, how an ideal value like beauty is unveiled through a complex web of real value qualities that are again tied back to objective dispositions in people and objects; that is, multiple value bearers (Figure 3.11). Observers – to come back to Plato – may talk about the "beauty" of a musical piece; but what they really refer to is what they are able to observe subsequently, which might be the pianist's performance or the harmony of the musical score. Or alternatively, they don't see the beauty of the piece at all, because the grand piano was out of tune. So, while humans frequently refer to ideal values in their language and are quick to judge on them, they should really humbly recognize that they always only perceive aspects of the ideal. Ideal values such as beauty are like the invisible vanishing point in a painting. Everything is ordered towards it, while it itself is invisible.[26]

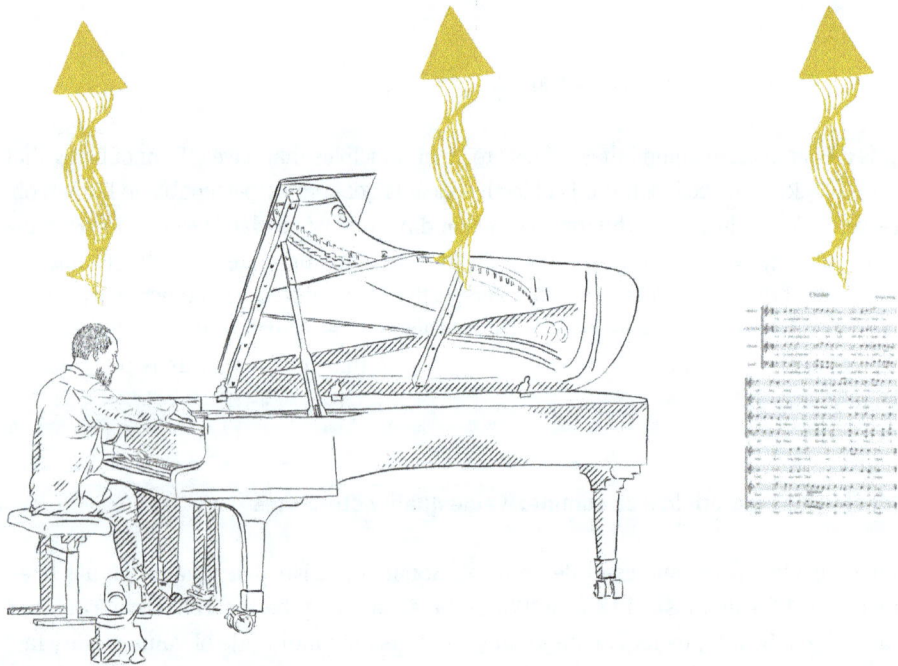

Figure 3.11: Multiple value bearers of beauty in a sonata.

The contextual meaning of values

Another aspect adding to the complexity of the value phenomenon is that a value, which we call by one name, such as beauty, means something completely different from one context to another. While it is possible to describe the beauty of a sonata with value qualities like the clarity, correctness, precision and gentleness of touch, the range and fullness of sound and harmony and the ease in listening, beauty in a person comes down to very different qualities. In his *History of Beauty*, Umberto Eco takes readers on a journey of how the ideal of human beauty has changed over the course of the past 2,500 years. He explains how human beauty was originally conceived of as symmetry of features and proportionality of limbs. These qualities were later complemented or marginalized by aspects such as naturalness (in the middle ages), or complemented by adornment and splendor in later times (Eco, 2004).[27] Today, beauty is regularly equated with slimness and athleticism (Figure 3.12). So, the real value qualities through which an ideal value like beauty manifests itself in the world always depends on context and historic time.

Figure 3.12: The qualities of the ideal value of beauty over time.

Conceptual value analysis

Even though the qualities of ideal values differ between value bearers, contexts and time, it is still true to say that ideal values also share some qualities by which they can repeatedly be recognized. There is an essence captured in words like "beauty" that is transferable from one context to another. For example, the qualities of symmetry, naturalness or splendor used to describe human beauty might as well be used to describe the qualities of a sonata. There is an essence in these qualities shared by the "concept" of beauty. Reflecting spontaneously about a sonata's qualities of beauty, concert visitors (stakeholders) might discuss the harmony and transport of the piece they listened to. But a music professional who knows about the structure of music might point out that the symmetry of a piece should also be considered in the valuation process. So, what is actually needed to understand an ideal value in depth is a conceptual analysis that brings together stakeholders' bottom-up quality observations with experts' top-down value quality knowledge. This is what takes place in VBE in order to understand the core value structures of a technology. Conceptual value analysis as a second-order analysis is used to systematically capture and complete what we can know about an ideal value's qualities.

Conceptual analysis of the value of privacy

A real-world example that shows how conceptual value analysis can enrich our understanding of an ideal value is a case study that was conducted with UNICEF in South Africa for a new IT platform called "Yoma." The goal of this project was to build a talent platform for African youth. Young people would find interesting projects through the platform and, upon participating in them, would build up an online CV with credentials garnered from the things they had done. They could curate their CVs through the system and also mentor others, once their CV and project history was above average.

Discussions with regional stakeholders found African youth to be concerned about a self-determined usage of their data. They would not want an unauthorized secondary use of their personal CV data. And they wished for their data to be securely stored and to not be inspectable by governments. The ideal core value of privacy as described bottom-up by stakeholders is depicted in Figure 3.13.

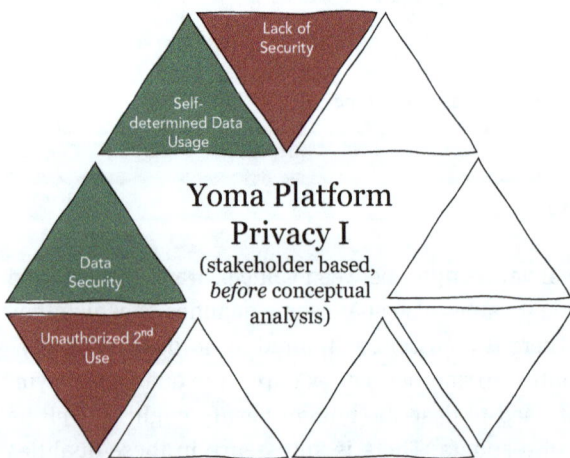

Figure 3.13: Bottom-up privacy (value) qualities seen by Yoma stakeholders.

What becomes apparent from looking at the value qualities derived from Yoma stakeholders is that only a few privacy qualities were progressively identified by them. A value lead and privacy expert would therefore need to combine this bottom-up conceptualization of the value of privacy with an official taxonomy of privacy, taken, for instance, from a legal source. The European Data Protection Regulation, for example, contains more privacy-related qualities relevant for Yoma. A conceptual analysis would add this top-down knowledge to derive an ideal understanding of privacy on the Yoma platform, adding such qualities as data processing transparency, data portability and accessibility, informed consent, etc. (see Figure 3.14).

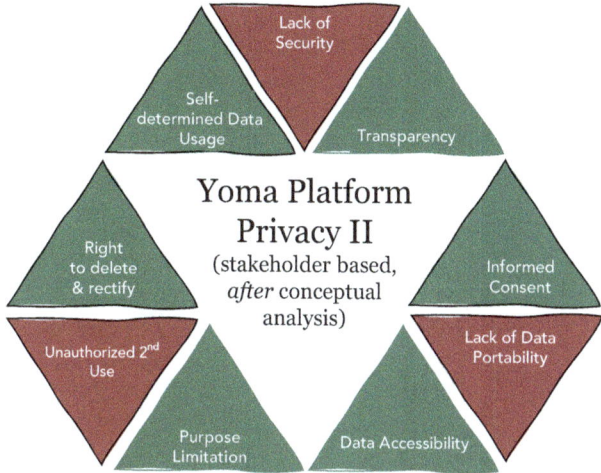

Figure 3.14: Yoma's privacy (value) quality cluster after top-down conceptual analysis.

What the two ideal value clusters (Figures 3.13 and 3.14) show is that the qualities through which an ideal value like privacy expresses itself in the real world come in very distinct forms and manifestations. They can come as enabling qualities, such as being informed about something; alternatively, they can come as something someone is able to do. Due to value qualities being so diverse, they are simply defined in the IEEE 7000™ standard as "potential manifestations of an ideal value, which are either instrumental to an ideal value or undermine it" (p. 23 in IEEE, 2021a). Note, though, that IEEE 7000™ calls value qualities "value demonstrators" (IEEE, 2021a). The term value demonstrator signals that an ideal value "demonstrates" itself in the world in a certain way; it shows itself; it becomes real, as Husserl would probably have said.

The two value quality clusters depicted in Figures 3.13 and 3.14 also show how ideal values are visually represented in VBE projects as being at the core of a cluster of value qualities (or demonstrators). This is why the IEEE 7000™ standardization project opted to call ideal values at the center of a cluster "core values." A core value is defined as "a value that is identified as central in the context of a system of interest" (p. 17 in IEEE, 2021a). In the following, the terms core value and ideal value are used interchangeably.

What should furthermore be noted is that the ideal core values at the center of clusters should normally be of *intrinsic* nature. Intrinsic values are those values that are valuable for their own sake, in themselves, on their own as an end (Ronnow-

Rassmussen, 2015). Examples for values of intrinsic nature are goodness, beauty, friendship, dignity, etc. No matter what culture one considers, all have a conception of intrinsic values, even though they might differ in the qualities by which they see these intrinsic values materialize. This is why they are ideal to put at the center of value clusters. All cultures can relate to them. The value qualities actualizing them, in contrast, are termed extrinsic values. *Extrinsic values* have been characterized as something valuable as a means, or for something else's sake (Ronnow-Rassmussen, 2015). Due to extrinsic values being a means for something else, they are also called "instrumental." In the security example given above, confidentiality, integrity and availability are instrumental to security. They are a means to the end of security. Therefore, they are extrinsic. We can apply these kinds of values back to the case of the piano-playing, where the intrinsic value of the beauty of the sonata depended on the extrinsic values of precision and gentleness of play.

Constitution of goodness through positive & negative values

The detailed analyses of the privacy and security values show that value qualities can be both positive and negative. Seeing one's own data being used for an undesired secondary purpose, for example, is negative. And, in such a moment, one could say that the positive value of privacy is suddenly non-existent. Its absence creates negative value. Following Franz Brentano, Max Scheler captured this dynamic in the following axiology (p. 82 in Scheler, 1921 (1973)):

- The existence of a positive value is itself a positive value.
- The non-existence of a positive value is itself a negative value.
- The existence of a negative value is itself a negative value.
- The non-existence of a negative value is itself a positive value.

Max Scheler did not discern the three layers of the value ontology as clearly as we have done it, distinguishing between ideal values, (real) value qualities and value dispositions. Furthermore, he also did not apply his philosophy to a system design context through which the existence of values can effectively be influenced. Refining Scheler's definition of goodness with a view to system design could read as follows:

- The existence or fostering of positive value qualities in a system constitutes positive value.
- The non-existence or undermining of positive value qualities in a system constitutes negative value.
- The existence or fostering of negative value qualities in a system constitutes negative value.
- The non-existence or prohibition of a negative value quality in a system is, in itself, a positive value.

Applying this value axiology to the practice of system design and user experience, it is in a company's interest to build all those value dispositions into a system that are needed to ensure not only the existence of positive value qualities, but also the exclusion of negative value qualities.

However, what happens if a system has so many value dispositions in place that only 80% of the value qualities can positively unfold? Will the ideal value still unveil itself to the user of the system, given that 20% of the value qualities potentially relevant for the ideal are missing? Here, we can turn back to the experience of values through progressive self-giving. The value of privacy will unveil itself up to the point where the observer suddenly recognizes that perhaps one value quality that she found important was absent. When this happens, she will doubt whether the overall ideal is being fulfilled. It is for this reason that innovating companies are well advised to test their early prototype systems thoroughly and monitor the unfolding of the value clusters in real-world deployment. Their aim should be to check whether relevant value qualities for the system and the context are absent, not perceived as they should be, or undermined. This practice of ensuring the existence of value qualities, and thereby, values, is also what links values to ethics. Max Scheler completed his axiology in the following way (p. 26 in Scheler, 1921 (1973)):

- Good is the value that is attached to the realization of a positive value in the sphere of willing.
- Evil is the value that is attached to the realization of a negative value in the sphere of willing.

Check questions

- What is the potential problem with a value definition that frames values as *conceptions of the desirable*?
- Why can the value judgments of one person be better than those of another person?
- How do we typically perceive values in our everyday life from a value ontological perspective?
- How do values relate to good and evil?
- How does value pluralism/relativism dissolve as an issue, once the three-layered value ontology is adopted?

Chapter 4
Value-Based Engineering phase 1: concept and context exploration

The goal of the first phase of VBE is to analyze and prepare the grounds for an optimal fit between a new system of interest (SOI) and the environment into which it is to be inserted. This is complemented by a feasibility analysis and the question of whether an organization is well advised to invest itself into the system.

What is being done here? The technology's conceptual vision is put into a high-level concept of operations for which the deployment conditions are investigated. Deployment conditions have a technical and a business side, both of which are, again, interwoven with social or ethical issues. On the technical side, it is established what component blocks and data flows the SOI has and by whom these are operated. In today's world of virtual integration of many distributed services into one customer-facing SOI, the partner network must be well understood and scrutinized. A first impression must be gained on what external partners exist to cooperate with for service delivery and under what conditions these partners work. Are they ready to share in an ethical service responsibility just as much as they are able to provide functionality? And, what are rules that can govern the application interfaces with them? Once these technical grounds have been surveyed, which is like investigating the soil and waters of a new garden and its surroundings (Figure 4.1), the question arises of who is impacted by the technology. Who are the visitors and keepers of the new technical garden? Who are the stakeholders relevant in the

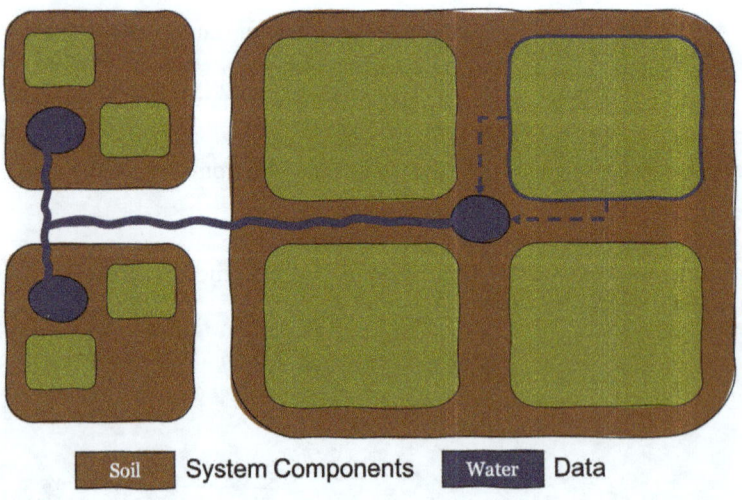

Figure 4.1: An SOI can be laid out as a garden.

SOI's context of envisioned deployment, and what is it that these stakeholders care about? These analyses allow us to conclude this first phase of VBE with a realistic assessment of the question: Is the system feasible?

The best moment to enter into this first phase of VBE is when there is a stated technological vision or a concrete technical capability already in place. In practice, this is a given, for instance, when lead users have come up with a prototype,[28] when Design Thinkers have come up with a first mock-up after product ideation,[29] or when startuppers have sketched out an early prototype and business canvas. That said, VBE is also usable in situations where a system of interest (SOI) is already deployed. But in such cases, the company that owns the SOI must be prepared to potentially re-engineer its entire technology, which can imply substantial financial investment.[30]

Seeking optimal fit with given conditions

In the first phase of VBE, stakeholders' values are not yet fully explored. However, in conversation with potential stakeholders and exploring their environment, it is natural that a first impression is gained on what value issues could play a role in a technology's future reception. For example, when analysts interviewed retail customers of futuristic test-stores in which RFID infrastructure was deployed, they immediately learned that people were worried about their purchases and belongings being chipped and tracked (Fusaro, 2004). They felt that their privacy could be invaded. They wondered whether RFID chips could be taken off after buying a product, or otherwise discarded. This stakeholder perspective, which was gained in the earliest technology planning phase, changed RFID chips' standard functional layout. A kill-function was foreseen in the concept of operation and the privacy issue noted as a potential legal or ethical feasibility issue in future deployments.

Naturally, the retail environment is only one of many specific contexts in which RFID technology might be employed. In other contexts, such as the use of RFID in hotel keys, a ripping of the chip or killing of the chip's read-out functionality would certainly not be desirable. This shows that depending on the context, one technology type might need a very different technical layout in order to meet stakeholder concerns than another one. For this reason, VBE requires project teams in this first phase of VBE analysis to discern the various contexts in which a technology might be deployed.[31] They are invited to physically or mentally explore this context to understand it at a deeper level.

Values versus needs

Note that in this exploration stakeholders are not asked about what they might "need" to improve their context conditions. Most of them won't know, for example, that they effectively need a kill function on a chip to protect their privacy. Instead, VBE tries to explore what it is that they would value about RFID's capability. One thing they might value is privacy. Another might be the seamlessness in doing some things with the help of the technology, like crossing borders or entering ski areas without queuing. With this focus on positive value, VBE differs from *needs*-based approaches. The latter often seem to imply that existing conditions are lacking something. Instead, VBE recognizes that people are normally satisfied and used to given conditions. At least in today's modern societies, they are not really lacking anything. It, therefore, might make us wonder whether and where stakeholders could see any additional value from a technology.

Technology can add a lot of value to the world, even where there is no substantial need. One may think about all the major innovations of these past decades. Did anyone need the Internet? Or a smartphone? Not really. But these technologies added a lot of value to societies, such as better access to knowledge, global integration, better reachability, etc. Some technologies may serve no greater purpose than simply being wonderful. Google Earth, for instance, originally did not do much more than letting people fly over the Earth, zoom into places of interest and view far-flung parts of the world from space. This is true value in the form of beauty and inspiration. Google Earth might have opened some peoples' consciousness to humbleness and respect for planet Earth. This kind of value proposition residing in mere beauty and inspiration for their own sake must not mean that business interests are neglected. In contrast, where there is human and social value, there will also be a willingness to pay. After all, the center of every business plan contains what is called a "value proposition" (Osterwalder & Pigneur, 2010), not a "needs-proposition".[32]

In some cases, of course, over-engineering or disruptive engineering (not infrequently inspired by bad management theory and practice) can lead to a destruction of the environment or a disruption of social structures. Technology is then called upon anew to help fix the negative value issues observed. This has happened a lot in the past 200 years – and in an accelerated form in the last 20 years. One may take as an example, the oft-criticized logistics operations of Amazon. As of 2021, the company has undoubtedly excelled in terms of classical process management values, which are cost, rapidity, flexibility and quality. However, the technological workflows used to create these process management values are only instrumental to financial gain. At the same time, they have a lot of negative value externalities that the stakeholders impacted by the technology complain about. For example, when Amazon's warehouse employees reported that they would not have enough time to go to the bathroom and are therefore forced to pee into empty bottles to ensure the rapidity of process flow, then their dignity is hurt (Liao, 2018). High organizational values that any organization should actually wish for, such as worker satisfaction (and thereby motivation), dignity and loyalty are undermined. VBE can help detect and fix such negative value quality potentials. But again, in its first phase, it primarily seeks to get an impression, lay out the system, talk to the people involved and get a feel for where, at the intersection between a new technology and humans, ethical feasibility might be undermined. The goal is to understand whether a good fit between technology and humans can be established.

Context fit and milieu acquaintance

Fit with the given conditions can only be understood if innovators (startuppers, product managers, system engineers, investors, etc.) invest time in delving deeply into the context and milieu of the space to which they want to add a new SOI. They must explore whether their technology idea – of which they only have a concept of operation – could work. Even though the monetary value proposition is often heralded, today, as being the main driver for market acceptance, the *compatibility* of a technology with its unique positive and negative value potentials should not be underestimated as a market success factor (Rogers, 1995). Take the example of driving a car with an automatic gearshift versus stick shifts. While in the USA almost 100% of cars sold are automatics, in Europe there are still roughly 40% of cars sold with stick shifts.[33] Attempts to promote automatics in Europe have been far less successful. Could it be a stronger desire for control that co-determines this European choice? Take another example, that of self-driving cars. While in the USA many roads are wide and spacey with limited traffic in dispersed countryside, European roads are narrow, twisty and busy. Will autonomous cars be as easily deployable in Europe as in the USA?

To explore such a priori questions, VBE starts with context exploration. It tries to understand, to return to the gardening metaphor, the weather and soil conditions as well as culture and purposes into which a new SOI is planned to be "planted." It does not make sense to plant palm trees, oranges and olives optimal for a fourfold garden in an Arabic town right into France's countryside, where it is rather potatoes and plums that grow in that climate (Figure 4.2). And, if contrary to common sense, you still do so, then it doesn't really fit and its maintenance is extremely expensive, which is the fate of so many ill-fitting IT systems today.

Startuppers who have professionally grown up in an industry into which they seed a new venture are well acquainted with the context. They are the ones who know the soil and climate milieu well. Where such experienced founders or domain experts are not part of the innovation team, leading management scholars like Ikujiro Nonaka recommend that innovators become industry apprentices first. Nonaka shared the (admittedly extreme) story, where an innovation team entered a real apprenticeship at a local bakery to learn the skill of how to treat dough by hand (see Chapter 8 for more detail). Only the tacit knowledge of how to treat the dough with a certain twist of hand allowed engineer apprentices to later build a superior bread-baking machine that was able to replicate this very subtle twist (see Chapter 7 for more detail).

64 — Chapter 4 Value-Based Engineering phase 1: concept and context exploration

Figure 4.2: The cultural context of deployment for an SOI is important for its future use.

When skilled innovators understand the "milieu" into which they insert a new technology, with all its routines, roles and intricate value qualities, they are in a superior position to create something valuable. In contrast, when new technology ventures do not know the milieu they are entering, they risk bringing something to market that is not compatible with reality, that creates limited competitive advantage or disrupts the status quo in a negative sense. As of 2021, about 90% of initially funded start-ups fail.[34] In all likelihood, this scrap rate could be dramatically reduced, if innovators were more acquainted with the domains they enter with a new technology.

In sum, VBE recommends first exploring the concept and context of a technology. Project teams depict, understand and analyze their SOI. They identify and understand the stakeholders affected. And, they test the (data) waters of their external partners and how these may impact stakeholders. For whom is it feasible to integrate into a value-based service? And who not? The first phase of VBE already sorts some wheat from the chaff.

SOI context exploration

The IEEE 7000^{TM} standard mandates companies to "describe the context of current operations to be replaced or changed by the future system" (p. 25 in IEEE, 2021a). This description should be done in a manner as free and unbiased as possible. Project teams should try to visit the context of deployment in person to grasp the (potentially various) local conditions where their future system will be rolled out. They should identify and talk to direct and indirect local stakeholders and start building a list of those who should be considered for further value exploration and ethical system design decisions. This start of a new innovation a project is similar to Design Thinking.

Design Thinking has become hugely successful in the first two decades of the new millennium, because it engages in exactly this context exploration (Brown, 2008). Design Thinkers "empathize" with direct stakeholders. They seek to understand existing routines and roles so deeply that they are able to uncover potential needs. The "ideation phase" of Design Thinking then focuses on improving the context investigated with a new product or service, addressing the deficiency, vacancy or inefficiency that might have been detected (Brown, 2008).

VBE deviates from Design Thinking only slightly: First, it always presumes some SOI (or, at least, a sketch thereof) as a starting point (Ahmed & Shepherd, 2012):[35] This is realistic. Companies might have obtained a patent, developed a new software capability, acquired IP or heard about a hyped technology that they feel is important to investigate for their own operations. For these reasons, some concept of operation or plan from a technology provider, etc. normally exists. Alternatively, a high-level Design Thinking project itself might have returned an early prototype for a system needed. In all these cases, an innovation team would have a first concept of operation at hand that can enter a VBE cycle.[36] Second, context exploration is sought not only with a view to direct stakeholders, but also indirect ones, such as society at large, nature or communities. Third, it is not assumed that stakeholders *need* anything; rather, the question is asked of whether the technology might contribute to uncovering a dormant value potential (see Chapter 7 for more detail). Direct stakeholder interviews can support a first insight into this question. Unlike Design Thinking, these stakeholders as well as the indirect ones are also explicitly asked about their technology fears and the negative potentials they see.

Envisioning deployment contexts that cannot be explored

Sometimes a real-world deployment context is not yet physically explorable. This can happen because the technology is too generic or because there is not yet any comparable service in the world. Something completely new might be offered, where nothing existed before – not even a field of practice. This was the case when the first smartphones, tablets, music players etc. came to market. A VBE example covered in this book is the African talent platform, Yoma, developed by UNICEF, for which there was no service precursor anywhere in the world. In such greenfield situations, it is helpful for innovation projects to envision the context of future system deployment with creative techniques and tools, such as those the Value Sensitive Design community has been putting forth. LeDantec et al. (2009) showed, for instance, how the ethnographic technique of photo elicitation could be of benefit to context exploration. Alternatively, science fiction scenes can be used; that is, a project team can develop visuals of what its concept of operations may look like to stakeholders, and these are then discussed or explored with a view to value potentials (for a critical discussion of science fiction, see Chapter 7). What is important in scenario analysis is that they are chosen in line with their ethical relevance. In other words: VBE seeks (and does not avoid) those contexts/scenarios or use cases for further analysis that may become an ethical challenge.

What if generic systems have no use context?

When asked to explore the deployment context(s) of an SOI, some organizations or engineering departments will argue that their systems are of such a generic nature that the context(s) of their later use is not yet known. For example, when computer vision algorithms are developed that translate optical signals into a precise picture representation, this kind of technology could be used pervasively. Computer vision algorithms can be used in contexts ranging from cancer recognition applications to military drone targeting systems. Therefore, engineers working on generic functionality focus mainly on measurable performance values such as efficiency or dependability for which they don't need concrete deployment contexts, except that they, sometimes, introduce the technical descriptions of their work with real-world examples. They stay at such a generic and fundamental level of technology emergence that the potential human, social or environmental value effects seem out of sight.

However, Google's "Project Maven" case, and indeed, many other historical cases (like the discovery of nuclear power), have demonstrated how engineers can be unpleasantly surprised when their generic technology or discoveries are used for purposes that they would never have wanted to support. In the Maven case, computer vision was suddenly put to use in drones for military purposes. Many engineers who had worked on the algorithms felt betrayed that their work effort, which was committed to good intentions, could be misused in such a way, aiding a military project (Makena, 2019). VBE, therefore, recommends considering context-of-use scenarios as early as possible in a technology's development lifecycle, and at the latest, when a system is trained for a known industry. In fact, there is a point relatively early in system design where a generic system is adapted to serve a final use. This is the point, for example, where the computer vision algorithm is trained with data from either a military or a health context. It is recommended to begin value-based analysis as soon as a technology is applied in this way to a concrete use-context.

System-of-System analysis

The exploration of context in VBE happens with some existing SOI in mind. While this SOI's form factor (e.g., user interface) might not be fully determined yet, there typically exists an initial technical sketch, or an early draft of what is called a "concept of operations." A concept of operations is a "verbal and/or graphic statement, in broad outline, of an organization's assumptions or intent in regard to an operation or series of operations" (p. 17 in IEEE, 2021a). It contains core system elements, stakeholders, data flows, and interfacing systems;[37] similar to a piece of land to be cultivated with a certain topography, water supply, interfacing neighbors and foreseen visitors.

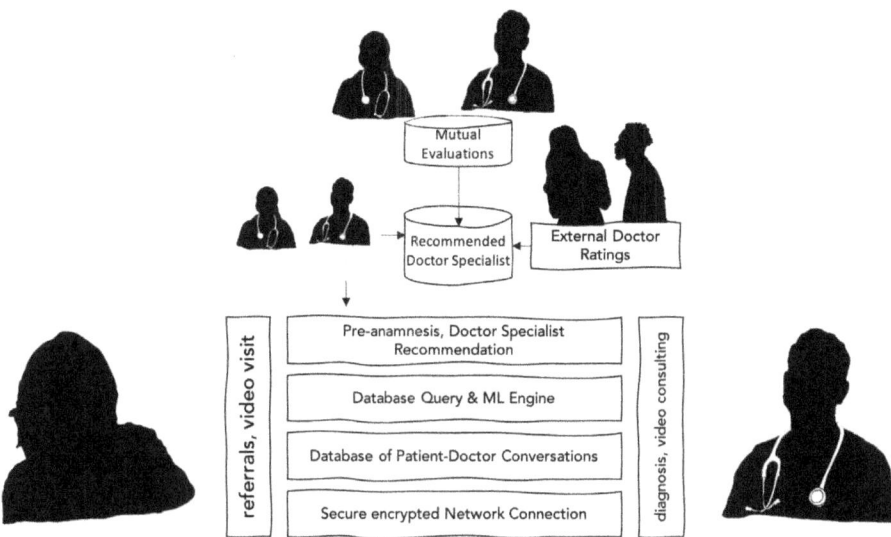

Figure 4.3: A block diagram can visualize the initial SOI setup.

A concept of operations can be captured with the help of different descriptions, depictions or modeling tools. It may start with a verbal description of purposes and properties, followed by a selection of use cases and scenarios, regional markets, stakeholders involved, etc. A simple block diagram may roughly depict the system elements of an SOI (Figure 4.3). Or, a UML component diagram may show, more precisely, the relationships (interactions) between relevant system nodes and their role in the overall architecture.[38] Between system elements, data flows should be captured, whereby anonymous versus personally identifiable data flows must be discerned as well as sensitive data flows. High-level Entity Relationship Models (ERMs) can detail data flows if needed, including data controllers and processors with their respective responsibilities (in line with the European Data Protection Regulation (European Commission, 2016).

If the SOI is a greenfield one, then the modeling at this stage must not necessarily be as detailed as many IT modeling languages allow for. For such a VBE SOI model, a high-level structured overview of the relevant technological components and data flows is enough, initially. However, if VBE is applied to already existing (brownfield) systems or analyzes the specific ethical challenges of a complex web of existing system components, then the models used to understand the SOI may be more detailed and complex.

Using context diagrams

One modeling approach to practically support the concept of operation description is a context diagram, which shows the external dependencies of an SOI (Figure 4.4). It is a graphical model that captures the data flows, and thereby, also the service flows occurring between an SOI and its environment.[39] These flows are especially important for understanding SOIs. As previously stated, data flows are like water for a garden. If organizations don't understand where their data-waters originate from, what contextual meaning the data has, what its syntax and semantics are, how reliably it flows, and how it is distributed and managed – then a system can degrade as quickly as a garden can.

Take the example of a telemedicine platform (abbreviated hereafter as "TM"). The TM start-up had the vision that patients would dial into its platform to speak via video with a general practitioner (doctor) and get a first diagnosis on a potential illness. TM doctors would give initial advice, write prescriptions and sick notes; but their principal task would consist of passing patients on to the right specialists for their need. For example, if a kidney problem is diagnosed, the TM doctor would access TM's database of kidney specialists in the patient's respective region and recommend not just any kidney specialist, but one particularly well-rated. The ratings in TM's specialist database would come from a network of recommending doctors that TM cooperates with (Figure 4.4). The context diagram visualizing this organizational idea shows that all operationally relevant interconnections of TM with its environment (adjunct IT systems) involve the transfer of sensitive personal health data. The diagram thereby signals how carefully data exchange must be organized. In such situations, the legal feasibility of the SOI might be at stake. The diagram also captures how many external partners are envisioned, forcing TM to depend on seamless unperturbed data exchange. The model gives an insight into the dependability and vulnerability of the company.

Generally, context diagrams allow project teams to engage in a first discussion on which system components are best for operating within one's own well controlled organizational boundaries and which ones might easily be outsourced to an SOS partner. Many companies, today, operate in an almost interwoven manner with their wider "system-of-systems" network (SOS). At their organizational boundaries, they interface with web services, databases and code components (third-party systems). Ian Sommerville defines an SOS as one that contains two or more independently managed elements (Sommerville, 2016). As the context diagram shows, TM planned to rely on a video chat application from an external video service provider. It wanted to store patients' health data with a remote cloud-service provider. It was thinking of using a potentially external pre-diagnosis AI service on top of its collected diagnosis data. An external mail service was considered for sending prescriptions, referrals and sick notes by mail to the patients. Thus, at least four external partner-systems fundamental to service delivery were initially considered for TM's SOI.

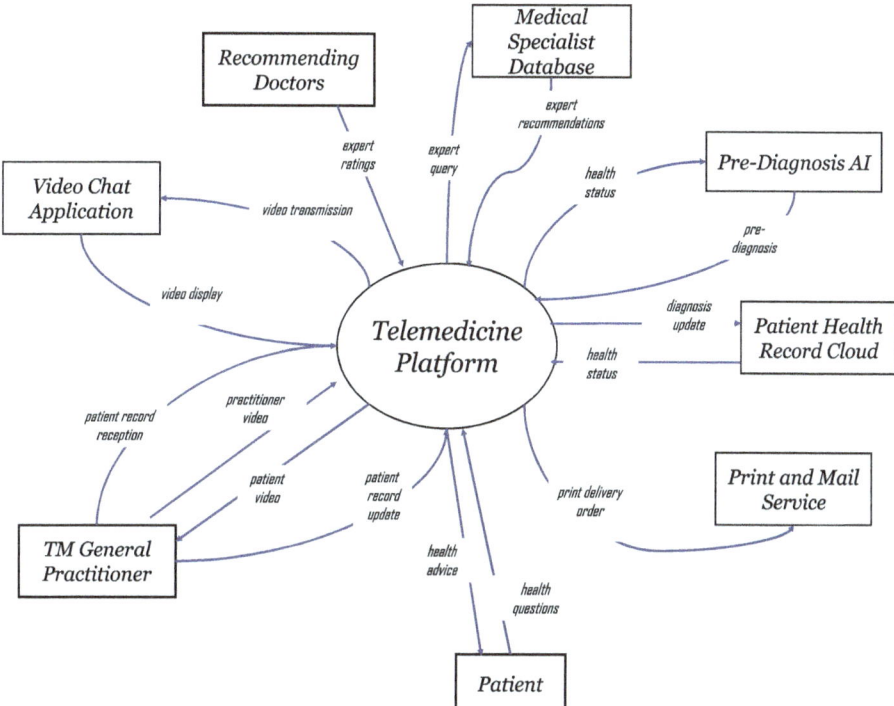

Figure 4.4: Contextual diagram of the TM Platform.

Why SOS analysis is important

Companies should be aware that a proper SOS analysis is in their own best interests in order to truly understand their value risks. All technically outsourced SOS service components ultimately present themselves as *one* system to customers. Customers will always regard the first-tier interaction partner or first-tier data controller as that entity, which is responsible for any ethical or value issues that arise; so even if TM might consider itself not responsible for its service partners, such as the video provider or the print and mail service, its customers will not see it that way. For this reason, the allocation decision between an organization's SOI and its external SOS entities should be a decision driven not only by cost or efficiency calculations. Instead, this allocation decision should be understood as deeply relevant for the quality of service that is ultimately delivered to customers; relevant, moreover, for the corporate brand, the organization's long-term market success, customer loyalty and the ability to effectively bear the responsibility that is ascribed by customers. Organizations that pursue VBE must know these dynamics – and, therefore, embrace responsibility for the value effects of their SOS ecosystem from the start of a project.

This extended responsibility for supply chain partners is captured through the arrows in the context diagram. These depict the direction of data exchange flows (dialogues) with partners as well as with users. Innovation teams can pick each external entity and reflect one by one on the legal, social and environmental feasibility of the partnership. What can go wrong? What are the challenges? What are the benefits? How much risk is involved? As soon as personal data is exchanged, ethical challenges associated with the privacy and security of the data are likely to arise; the challenges only mount if the personal data exchanged is sensitive, not collected legitimately or comprises dubious quality levels.

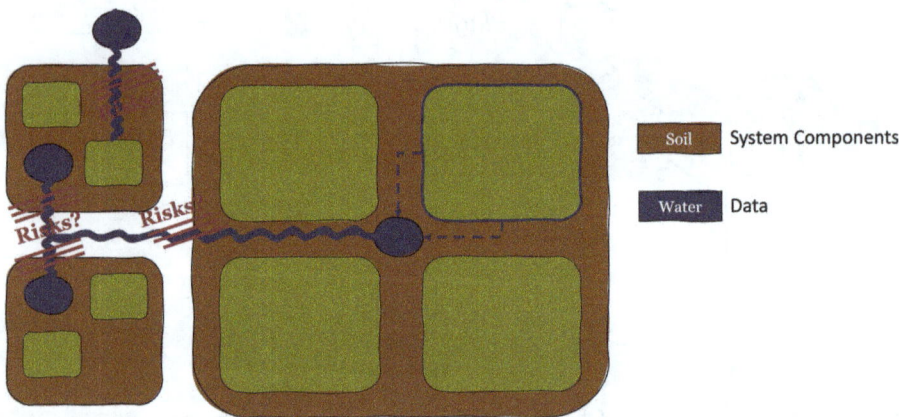

Figure 4.5: Some parts of an SOI will bear risks.

However, it is not only data issues that should be analyzed at the interface between two organizational entities. It is also the nature, background and general risk associated with partners that an ethical value-based organization should reflect on (Figure 4.5). In one VBE project conducted in Africa, for example, it was found that one of the early envisioned SOS partners had a potentially questionable apartheid heritage. As a result, VBE would recommend double-checking the operations of that partner. "Show me your friends, and I will tell you who you are," as the old saying goes.

Partner analysis

Analyzing one's SOS partners (their data use conditions, corporate, nature, background, philosophy and data operations) requires research and background checks. And such analysis alone is not enough. VBE requires organizations to also have sufficient control over the partners they pick. One crucial activity required by the IEEE 7000™ standard specifies that the SOI organization needs to "obtain access to the enabling systems or services to be used" (p. 38). This implies that it is not recommended for VBE organizations to just interface with commercial off-the-shelf services (COTs) or any kind of black box component. Instead, they should work towards "acknowledged" or "directed" forms of partnership (ISO, 2015) (Figure 4.6). Acknowledged forms of SOS partnership integrate independently owned and operated system elements only where the operators have a service level agreement (SLA) and a designated management that is responsible for the interface between the services. Terms and conditions of accessibility to the partner's operations can be specified in the SLA. Alternatively, two partners can decide to collaborate on a joint service, which they integrate and build together for the respective service vision at hand. This is a directed form of partnership. Here, they can ensure observability and control over likely ethical issues, from the start.

Type of SOS	Relationship Character as described in ISO/IEC/IEEE 15288-2015	Observability of ethical issues	Control over ethical issues
Virtual systems	– Lack of central management authority – Lack of centrally agreed upon purpose – Emerging behaviors that rely upon relatively invisible mechanisms to maintain it	None/ very low	None/ very low
Collaborative systems	– Component systems interact voluntarily to fulfill agreed upon purposes; collectively decide how to interoperate, enforcing and maintaining standards	low	low
Acknowledged systems	– Recognize objectives, a designated manager and resources for the SoS – Constituent systems retain their independent ownership, management and resources	medium	medium
Directed systems	– Integrated SoS built and managed to fulfill specific purposes – Centrally managed and evolved – Component systems maintain ability to operate independently – Normal operational mode is subordinated to a central purpose	high	high

Figure 4.6: Types of system of system partnership form (p. 67 in IEEE, 2021a).

That said, many of today's systems are, in actuality, built with the help of less controllable external service components. These are called "collaborative" or "virtual" partnerships. An example of such a collaborative SOS is when a payment service component is integrated in an online retailer's website. Even a small online retailer can offer its customers credit card processing in this way, without doing the clearing and processing itself. Another example is cloud service provision, which ensures that even small websites and services can rapidly scale their operations, without running into hardware or processing limits. When it comes to such well-matured and ubiquitously used service components as payment or processing, it is difficult for an SOI organization to forgo partnering with big COT services and insist, instead, on an acknowledged or directed form of partnership, in order to act with ethical responsibility." It is for this reason that IEEE 7000^{TM} does not mandate acknowledged or directed forms of SOS collaboration, even though it recommends them. Figure 4.6 summarizes the various forms of SOS partnerships as they are recognized in ISO 15288 (ISO, 2015) and extended in IEEE 7000^{TM}'s Appendix E (p. 66 seq in IEEE, 2021a).

Ensuring control over remote AI components

One special form of partnership is interfacing with an AI-based service component, whereby an external "skill," "intelligence" or "foundation model" is integrated into an SOI (do you recognized the AI in Figure 4.7?). Systems can be associated with the term "Artificial Intelligence" when sophisticated data processing techniques such as machine learning are used, but also when more traditional reasoning techniques are applied to Big Data volumes.[40]

For an external AI service to be integrable in a trustworthy manner, the IEEE 7000^{TM} standard outlines that control over the following should be in place:
- the quality of the data used in the AI system;
- the selection processes feeding the AI;
- the algorithm design;
- the evolution of the AI's logic; and
- the best available techniques (BATs) for a sufficient level of transparency on how the AI is learning and reaching its conclusions. (p. 68 in IEEE, 2021a).

Figure 4.7: Complex SOS environments can contain AI services that are themselves interlinked with other services.

Challenges of concept and context exploration

Drawing the right system boundaries

A challenge in analyzing partner systems is that organizations can ignore some important connections (interfaces or edges) of their SOI to relevant partners or services. They can misjudge or under-represent their true system boundaries (Figure 4.8). Take the case of the retailer described above, who wanted to introduce RFID on its shop floors and, in this vein, needed to prove the privacy-friendliness of its RFID-based retail checkout cashiers. The company argued that the checkout cashiers physically deployed in their supermarkets would have no privacy issues, since they just used RFID readers to locally read out customer products and calculate a bill. No personal information would be involved in such decentralized billing transactions at the supermarket. What the retailer ignored in this analysis was the interaction of these checkout systems with a national loyalty program the company participated in. With the remote loyalty program, each customer could later be re-identified in the loyalty program's database, causing privacy risks. This example makes plain that organizations can be tempted to include only those system elements in a value analysis where they already intuit that these are ethically and legally unproblematic. Or they simply underestimate the complexity and risky parts of their own operations. VBE encourages organizations to face their complex interaction challenges and rigorously scrutinize all data exchanges happening, particularly those where personally identifiable data could be involved.

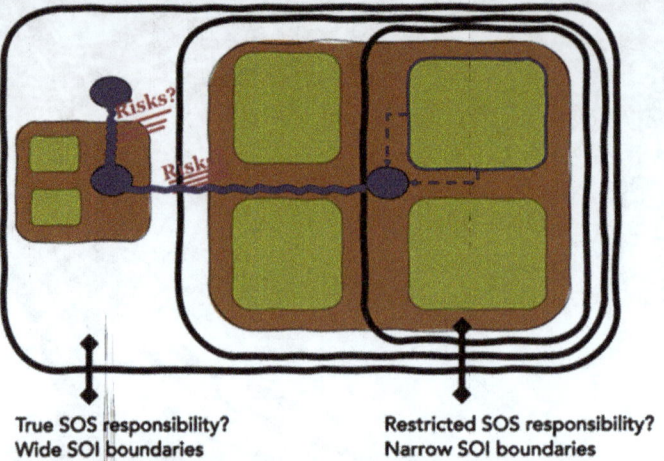

True SOS responsibility?
Wide SOI boundaries

Restricted SOS responsibility?
Narrow SOI boundaries

Figure 4.8: The analysis boundaries of an SOI with its SOS need to be determined.

Handling the complexity of the SOI and the SOS

Another challenge that can arise is the sheer complexity of an SOI. Sometimes, there are so many external elements and partners involved that a company engaging in VBE does not know where to start and what level of detail to look at, for each partner. In this case, it is not recommended to explore ethical issues via a single view or perspective on the SOI. What is instead necessary, is to split the concept of operation up into multiple meaningful parts; that is, to effectively create a number of distinct "SOI views" that are then analyzed separately (Figure 4.9).

A separate SOI view for ethical analysis is important for those SOS partner structures that have a potentially massive influence on the ethical nature of an SOI (its safety, security, reliability, trustworthiness, etc.). If an SOS partner, X, supplied, for instance, a large part of the data that the SOI under scrutiny needs in order to offer a service, and if this external data needed to have a specific quality level, then the relationship between this partner, X and the SOI organization would deserve its own analysis through a separate SOI view. In such a separate SOI view, a VBE organization would seek to understand the particular operations of its partner, X, and scrutinize its specific conduct in order to understand the potential ethical implications of that particular partnership. This could include the scrutiny of data collection methods, data quality levels, data usage policies, etc.

The African talent platform can serve as an example to illustrate the importance of SOI views. The initial concept of operation foresaw data pooling and intelligence sharing with three partners, A, B and C. A was a learning platform engaging Africans in online learning-challenges, B, a survey-tool asking young Africans questions on regional issues and C was a reward-system, distributing reward points for small work tasks that could be exchanged for food or transport services. VBE analysis needed three separate SOI views in this case, one for each of the partners, since the learning platform partner, A, seemed to engage in rather shallow rating and ranking of its users, and there was, therefore, a risk that it might supply questionable data quality to the SOI. The envisioned partner, B, provided data on extremely sensitive issues like genital mutilation or a regional municipality's conduct – information that understandably needed a high level of security protection. And, partner C had the apartheid background already mentioned above that would need to be checked out in more depth. With each partnership, the ethical SOI's service vision was, therefore, prone to run into unique ethical risks. For this reason, VBE analysis was split into three separate SOI views for value analysis, with separate partner stakeholders on board.

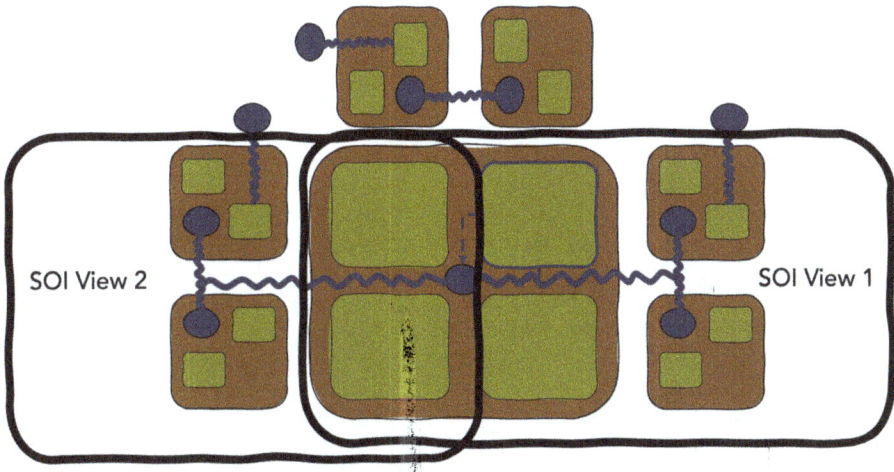

Figure 4.9: Different SOI Views facilitate VBE analysis.

Stakeholder identification and roles

Once the broad concept of operations, the system boundaries and partners are clear, it is possible to identify relevant stakeholders (Figure 4.10). Stakeholders are individuals, organizations, groups or other entities that can affect, be affected or perceive themselves to be affected by an SOI. They "have a legitimate right, share, claim, influence or interest in a system" (p. 10 in ISO, 2015). Two stakeholder categories should be discerned: First, those who directly interact with a technology, such as human beings directly using the system (end users) or organizations purchasing the system (acquirers). And second, those "stakeholders who, although they never or rarely interact with the system as end users, are nevertheless affected by the system" (p. 38 in Friedman & Hendry, 2019). This latter group is called "indirect stakeholders." Examples include communities, neighborhoods, institutions, nation states and future generations – as also animals, nature or entities with historic or sacred meaning.

VBE requires that all of these stakeholders are identified and that relevant and suitable representatives are appointed to stand in for these, through every single phase of further system value exploration and system design. Stakeholders are not bundled into one homogeneous group called "users." Instead they are acknowledged in their very specific roles, similar to what usability researchers refer to as "personas" (Pruitt & Grudin, 2003).[41] Since an SOI (or rather, SOS) normally affects people in so many different roles, only an extensive and diverse group of stakeholder representatives has a chance to anticipate a reliably complete value spectrum for the SOI. Stakeholder representatives should sensitize for minorities and be critical of the SOI. Any international rollout of technology should be accompanied by the inclusion of stakeholder representatives stemming from those regions of the world in which a system will be deployed (Figure 4.10).

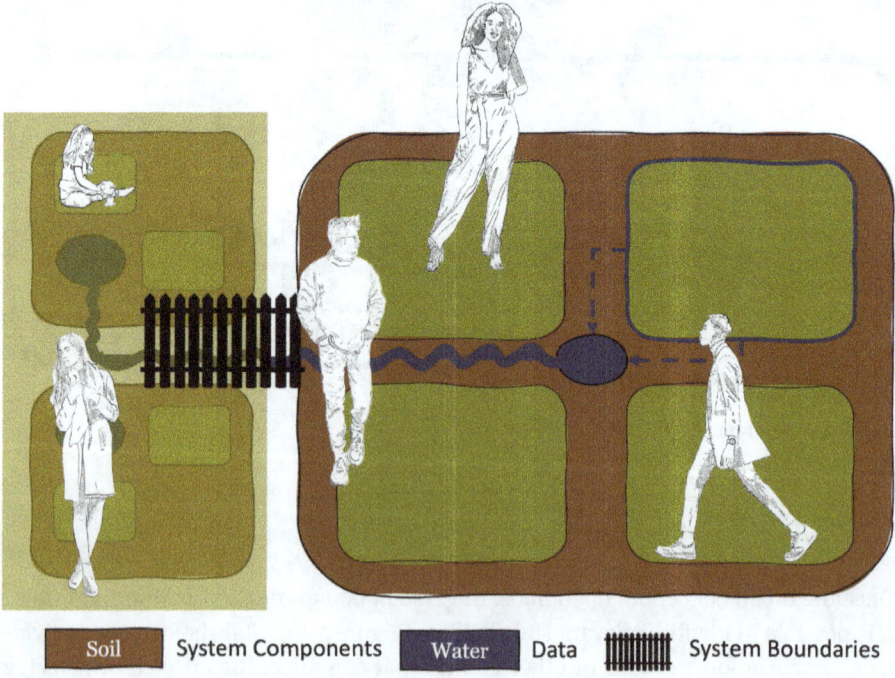

Figure 4.10: A system has direct and indirect stakeholders to its various parts.

In the TM case, we identified 20 such varying stakeholder roles. For example, among patient users we discerned student patients who might seek quick and efficient sick leaves on exam days, elderly patients who might benefit from the remote service but did not have Internet access, or foreign patients who didn't speak the national language or might not be insured (but still need access to health services). Indirect stakeholder roles included young doctors whose reputation was, as yet, insufficient to garner a good rating, or the doctoral community as such having a view on the trend of doctoral peer-rating. In the latter case, the stakeholder was not a person, but an abstract entity; namely, a community affected by the SOI.

For the corporate context, Ulrich, (2000) showed that the sources of stakeholder motivation, power and knowledge, as well as their legitimation should also be considered in their selection (Figure 4.11). Senior managers, for instance, are often pressured to focus strictly on revenue and profits instead of an ethical and value-inspired design. They may press for cost reduction through IT, low-cost solutions, automation (job replacement), or other ethically ambiguous goals. Alongside such varying

Figure 4.11: Stakeholder dialogue needs discourse ethics.

economic motivations, stakeholders bring different worldviews to the table. Attributes such as their personal values, political attitudes, virtue capacity, intellectual strength, and farsightedness vary. On top of this individual diversity, organizational politicking is a common practice. "Politicking" refers not only to personal power plays but also to the fact that stakeholders may represent different interest groups (such as company shareholders or workers) when they discuss the deployment of a new IT system (for more detail, see pp. 173–176 in Spiekermann, 2016). In VBE projects, it is important that such dynamics are understood and that all parties have a respected place at the table.

Ensuring ideal speech situations

Due to the diverse stakeholder backgrounds and potential politicking, VBE and IEEE 7000™ encourage ideal speech situations (Habermas, 1985) when stakeholders meet. This means that project conditions are created from the start, where stakeholder representatives can effectively bring in their own true voice. (Mingers & Walsham, 2010) have described the traits of ideal speech situations. According to them, stakeholders should be allowed equal participation, be encouraged to question claims and assertions and be able to freely express attitudes, desires and needs (for more detail, see pp. 173–176 in Spiekermann, 2016). In the IEEE 7000™ Transparency Management Process for VBE, organizational rules for transparency and communication are enforced. It is noted that arguments must be truthful, factually correct, intelligible and sincere (p. 50 in IEEE, 2021a).

Recognizing the socio-technical nature of an SOI

Against the background of stakeholder considerations and extended SOS analysis it is clear that the understanding of the word "system" in Value-Based Engineering and IEEE 7000™ is a socio-technical one. While a context diagram would only depict an initial technical context view, wider stakeholder involvement ensures that ethical systems embed the concerns of society at large. Socio-technical systems regard technology as embedded in organizational public or private processes, which include policies, people, preferences and incentive systems (Mumford, 2000).

Feasibility analysis

As innovation projects run through the various analyses of the SOI, examining its context, its interfaces, interfacing partner services and stakeholders, they subsequently gain an understanding of the SOI's feasibility. A structured and founded judgment emerges on whether it makes sense to pursue a technology idea or not.

Hence, each of the four detailed analyses in this VBE phase (Figure 4.12) should be accompanied by notes on what it implies for system feasibility.

Feasibility analyses can have multiple assessment dimensions (Hoffer, George, & Valacich, 2002; Spiekermann, 2016), including:
- a legal assessment of potential legal or contractual ramifications of the SOI
- a political assessment of how key stakeholders in the organization view the SOI;
- a technical assessment of the development organization's ability to construct the proposed SOI;
- an operational assessment of whether and how the SOI will fit into existing operations

For VBE projects, the ethical feasibility analysis is added to this list:
- Ethical feasibility looks into first value issues for affected stakeholders.

Each of these feasibility assessment angles encourages a project team to think about the SOI from a fresh perspective. And, every assessment can lead to further refinements and adjustment of the concept of operation, which, again, has implications for the cost and complexity of the project, hence the business case. This is why VBE's first phase should ideally be aligned with business case construction, where a realistic estimate can be made regarding real project risks. An economic assessment of the real financial benefits and costs of an SOI can be added here, as well as an assessment of the time schedule: Can the SOI project be completed in the

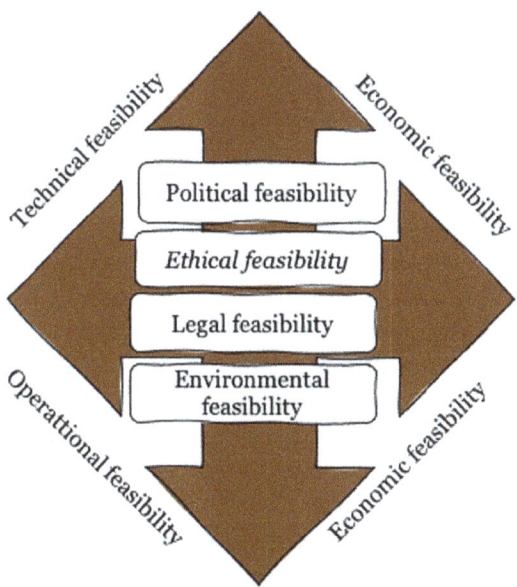

Figure 4.12: Dimensions of an SOI's feasibility.

necessary time frame to match organizational goals and address the value potentials and challenges likely to arise?

Questioning an SOI's feasibility

Political, legal and ethical feasibility are core dimensions for VBE (incorporating the social and environmental feasibility detailed in IEEE 7000™). Testing the political feasibility of a new system means evaluating whether key stakeholders will support it. A system may, for instance, affect employees' job satisfaction or the satisfaction of those expected to build and market the SOI. More broadly, all relevant stakeholders interviewed in the deployment context(s) will naturally have a perspective on the SOI. Note, however, that political feasibility is not necessarily the same as ethical feasibility. Actions that are politically correct and receive stakeholder buy-in are not automatically good. Political feasibility means that stakeholders *accept* the SOI and its implications. They believe that they can live with it. But not everything that is politically accepted and even economically and legally doable necessarily generates positive value, is ethically tenable or is even appreciated. Take again, the example of the logistics processes at Amazon. Of course, these must have seemed politically feasible when the company first introduced them. They certainly made economic sense. And, presumably, a professional company like Amazon would have checked on the legal feasibility of its workflow systems before implementing them. However, as the ethical issues arise (in this case, around worker treatment) motivation drops, whistleblowers pop up and policy makers observe the ethical challenges of such systems. As a result, repercussions emerge. The lack of ethical feasibility is addressed by new laws and regulations that question the political feasibility in the aftermath. In the end, a company like Amazon might be forced to re-engineer its entire workflow system to align it with stakeholder interests.

Re-engineering becomes necessary, for instance, in the context of privacy and data protection regulation. As personal data markets have started to thrive and personal data is heralded to be the "oil" of the digital economy, the temptation grows for companies to use personal data collected from customers for clandestine purposes. As of 2022, secondary use of data has become a major source of income for many online companies. Typically, companies require customers to consent to data sharing practices by signing terms and conditions that are incomprehensible. The companies then generate revenue from those users who effectively don't read the terms but still sign them. In this example, economic feasibility is a given. Legally, the practice is argued to be largely feasible. However, global surveys show that 80–90% of people worry about such practices and want more control over their data.[42] Therefore, ethically speaking, these practices have been questioned, and the systems enabling them are candidates for re-engineering as legal sanctions grow increasingly strict.

VBE tries to avoid such conflicts and re-engineering expenses by integrating ethical feasibility analysis, from the start. One of its most important contributions lies in the identification of potential negative value externalities in the early phases of system design. This is done in detail in the value exploration phase, described in the next chapter. But even in the earliest stages of concept and context exploration, ethical feasibility is interrogated and noted. A good test of ethical feasibility is to ask project teams elaborating the operation concept whether they would be ready to publicize the details of an SOI's workings online for NGO scrutiny. If a company can answer this question affirmatively, it is on a good path.

Consequences of context and concept exploration

Reflecting on these analyses, several observations can be made:

First, the concept of operations is typically adjusted in this first phase of VBE. Organizations realize that their initial business and service idea may not exactly be the one that makes most sense for the context explored. As a result, they may adjust the mission of the SOI. Organizations may, in doing so, also abandon some of their service partners. It may also be that some partners will simply be replaced by more suitable ones, or partners will be consulted to jointly agree on service quality. Alternatively, the organization may decide to build and provide some of the service elements in-house that it had earlier planned to outsource. In other words, the result is the development of an "alternative concept of operations."[43]

Second, an organization might feel that its SOI is not ethically sensitive at all. The concept and context exploration suggests that further detailed analysis of ethical issues and values followed by an ethically aligned design is simply not necessary.

Third, and this is a point specific to Value-Based Engineering and not covered in IEEE 7000™: Organizations should reflect on whether they want to build and/or invest in a system at all. Innovation management scholars agree that innovation project selection should always be a funnel, not a tunnel (Cooper, 2008). Just because an organization has an idea and has explored a concept of operation, it does not mean that it should invest in that system and pursue it further. It can and should be willing to step back from an idea, because it found in the first phase of concept exploration that there were too many challenges and issues to navigate.

Check questions

- How can and should an SOI be modeled for further VBE analysis?
- What forms of SOS partnerships can be discerned, and which ones are advisable for VBE projects?
- What kind of interface responsibilities exists if an SOI uses external AI components?
- Why can it be challenging to pick the right system boundaries?
- Why are different SOI views potentially needed?

Chapter 5
Value-Based Engineering phase 2: value exploration

The goal of the second phase of VBE is to explore the positive and negative value potentials borne by an SOI. Since the SOI, with all its context factors and stakeholders, has been fathomed, it is possible to reflect on the positive values it could bring into the world to enrich stakeholders. And furthermore, the negative values it must prohibit from unfolding through its design. The answers to these two questions are prepared with the help of moral philosophy, followed by value prioritization and conceptualization (Figure 5.1).

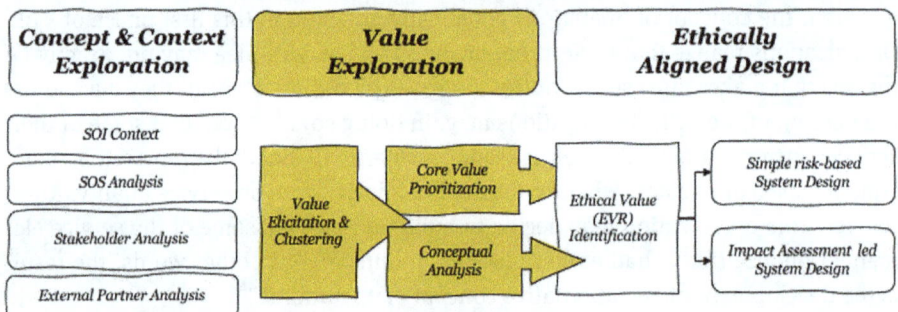

Figure 5.1: The value exploration phase in the VBE process.

Ethics is not (only) about morality

When learning that VBE with IEEE 7000™ implies applying moral philosophy to a concrete technology, it appears as if VBE is about corporate morality. The framing of IEEE 7000™ as an "ethical model process" further supports this notion. Consequently, executives could feel cornered into behaving in a specific way. There is, however, a difference between morality and ethics. Morality has traditionally been defined with a view to human behavior, mainly in terms of right, honest and acceptable behavior, as judged by cultural norms (Cambridge Dictionary, 2014). Ethics, in contrast, is not limited to people's behavior. As a theory, it can also apply to artifacts, symbols, relationships, etc. (Scheler, 1921 (1973)). In our modern understanding, ethics is a wider concept than morality. This is important in VBE, which deals primarily with machines.[44] Project teams are encouraged to anticipate and assess the positive and negative value potentials of a machine, which initially hasn't much to do with corporate morality. And even if the prioritization of values for system design later needs moral guidance, the approach as a whole does not tell organizations what is right or wrong. It just helps organizations to think about what is good or bad.

This thinking about what is good for an organization and how this goodness could be undermined (by a bad system) is of a higher nature than the question of righteousness (treated in moral philosophy), which is subsidiary to it.[45]

Against this background, ethics is defined in VBE as a branch of knowledge or theory that investigates the correct reasons for thinking that this or that is good.[46] And, these "correct reasons" are anchored in values. A good system is one that is engineered in such a way that its value dispositions trigger the unfolding of relevant positive value qualities and prohibit the unfolding of negative value qualities. In line with this understanding, IEEE 7000™ defines the word "ethical" as "supporting the realization of positive values or the reduction of negative values" (p. 18 in IEEE, 2021a). An example is an AI system that actualizes the positive value of transparency and prohibits a lack of privacy or security.

Setting the scene for value exploration

The first step of value exploration is value elicitation. Value elicitation depends on a sound understanding of the SOI, partner services and stakeholders. Take note of this starting point for value analysis: Project teams (including stakeholder representatives) scrutinize a concrete SOI (or several SOI views) that have technical depictions, system elements, data flows, some already existing roadmap, etc. The project team's creative anchor is hence not some generic technology, a broad future-world scenario or a narrative of the world in which a respective technology can go wrong; it is always a corporate, socio-technical depiction – ideally, one for which the (future) deployment context has already been visited.

This starting point is relevant because if ethical discussions are not practically bound to a concrete SOI but rather to some general science fiction narrative, then they can easily go astray. Take the TM case described earlier. A future narrative of telemedicine, in general, might suggest that there is going to be a virtual doctor-bot speaking to a patient. The AI-based diagnosis would be done completely virtually. A virtual doctor-bot is probably cheaper than a real one. Since it possesses "artificial" intelligence instead of "human" intelligence, many people these days would trust the doctor-bot more than a human doctor. As a result, an ethical analysis might pivot around issues like labor market problems for future doctors and weighing the costs for fictitious doctor-bots versus those of real doctors. Furthermore, what a virtual doctor-bot should look like, whether it should anthropomorphize, etc. might be up for debate. This kind of discussion, however, does not help a telemedicine start-up in its endeavor to solve concrete value issues at hand with their telemedicine system, their partner network and their local customers. Looking at a concrete SOI here and now with SOS partners who provide a specific and probably limited set of functional qualities, forces a project to focus on what is relevant: the value-reality of the SOI at hand.

That said, two elements of imagination are still important when values are explored for a concrete SOI: time and scale. When the roll-out effects of an SOI are envisioned, VBE assumes that it will be used in the next 10 years at least and at scale, or "pervasively", as Friedman and Hendry (2012a) term it. Some technologies, of course, do not need scale to do harm or to contribute a positive value. But VBE still assumes scale, because in the past 30 years, the digital economy has observed a dynamic called the "network effects," which entails systems that sometimes attract millions of users in just a few years, months or even days. In fact, this is what many start-ups strive for. But it is often due to this unexpected and rapid scaling that the ethical issues of a service are felt by the society. Take the example of the social network Facebook. The allegations Facebook has faced in recent years, such as the charge of undermining democratic elections, rose to public awareness because of Facebook's unique position in the market. If Facebook only served 500 customers in Ohio, societies worldwide would undoubtedly care less about its ability to influence elections. Many ethical issues only become relevant through scale.[47]

Including stakeholder representatives

The first pillar of successful value exploration is the involvment of direct and indirect stakeholder representatives. Stakeholder representatives should be regarded as stewards entrusted with the responsibility of voicing the interests of those who cannot participate in the innovation project. They should have the credentials to speak for the interest group(s), person(s) or the entity they represent. Examples are non-governmental agency (NGO) representatives, direct end-user/customer representatives, union representatives, etc.

What is important in picking these stewards is to ensure that critical thinkers are on board. As Werner Ulrich has shown, appropriate motivation, power, knowledge, and legitimacy among stakeholder representatives are crucial for eliciting the true ethical challenges of a project (Ulrich, 2000).

The chosen stakeholder representatives in this and in the other VBE phases are (at least in part) recurring members of the innovation project team. Value exploration can engage an extensive number (perhaps 8–12) of people. In the third VBE phase where the SOI is designed (Chapter 6), some of these continue to be part of a smaller stakeholder group, verifying that the values explored and prioritized effectively end up in the system. This latter stakeholder group should be seen as an addition to the technical system engineering expert(s) and the product manager(s), who are otherwise responsible for the SOI's development.

Appointing Value Leads

The second pillar of success for value exploration is the appointment of a competent "Value Lead." A Value Lead is defined in IEEE 7000™ as a "person assigned to coordinate and conduct tasks related to ethical value elicitation and prioritization and traceability of values through the requirements and design artefacts" (p. 23 in IEEE, 2021a) (Figure 5.2). So far, companies have no tradition of hiring people in a position called "Value Lead," even though there have been some provisional discussions on whether big corporations should have a "Chief Value Officer." When there are value issues arising in projects today, these are often delegated to legal or CSR departments that are supposed to take care of them (Bednar, Spiekermann, & Langheinrich, 2019; Lahlou, Langheinrich, & Röcker, 2005). This "offloading" of ethics has not necessarily been due to engineers' lack of interest in the matter; rather, organizations tend to not give engineering staff the necessary time, autonomy and resources to embrace ethical responsibility. Hence, it is a matter of engineers' personal motivation to work after-hours and deliver ethical components, in addition to the paid-for project scope (Spiekermann et al., 2018). If such a personal commitment is not given, ethical requirements easily fall off the end of the long list of a system's requirements. Ethical issues are "swept under the rug", as former Siemens engineer Brian Berenbach put it (Berenbach & Broy, 2009).

Figure 5.2: Value Lead at work.

This aspect changes with VBE and when IEEE 7000™ is used. Here, Value Leads are part of the project team and help it to run through value exploration and later monitor as well as document that the system design is ethically aligned with the values that are found to be relevant. Value Leads ensure that the risk of identified values being undermined is minimized and that the positive value potentials are consistently fostered through system design. In short, Value Leads drive the value mission. They are the ones who elicit from stakeholders, what values they deem relevant in a system. They bring the moral frameworks to the project team that facilitate and morally legitimize the value elicitation process. They cluster the elicited values and then help organizational leaders to prioritize them. They conceptually refine the prioritized value clusters and facilitate the identification of ethical value requirements (EVRs). Subsequently, they support the derivation of system requirements and ensure that these requirements end up in the system. Finally, they monitor the success of the system and analyze whether values unfold as expected. If not, they have to trigger a new iteration of the VBE process in order to mitigate the unexpected unfolding of negative values. Alternatively, they trigger a further strengthening of any positive values that are not materializing as planned.

Some companies might be tempted to delegate the role of Value Lead to someone outside of the core innovation project; for instance, to a representative from the CSR department, someone from the worker's union or to an external consultant. Note, though, that IEEE 7000™ outlines that "the Value Lead is not 'the person in charge of ethics' in a project . . ." Even if the Value Lead is involved in many tasks for the project, she or he only "contributes subject matter expertise and facilitative skills, bridging gaps between engineering, management, and human values in a constructive way" (p. 22). The reason why this is noted in the 7000 standard is to clarify that organizations should not offload the value work to any one person or organizational unit; and certainly not to someone outside of the core innovation team. Instead, all parties involved in a project must participate in the value mission definition and realization. The Value Lead should, therefore, be an employee close to the SOI's development, who is foreseen to hold a permanent position involved in managing it (for example, a product manager or a system engineer responsible for the SOI).

Against the background of these tasks and roles, Value Leads need to be well-qualified employees. They need to possess adequate knowledge on what values are as well as the moral theories that allow for their elicitation. They should have the knowledge to conceptually analyze values and be acquainted with the technical and organizational requirements that enable them. They should be acquainted with internationally agreed value principal lists (human rights agreements, legislation, industry principles, etc.). Their conceptual, verbal and communicative skills should be sophisticated. At the same time, they must also be able to relate to the technical system elements and understand technical requirements at various levels of system design and architecture. They should know risk management work and how it can be used to mitigate system threats.

The philosophical foundations of value elicitation

With the right people on board, budget and time granted, guidance from an assigned Value Lead, having all project participants informed about the SOI's concept of operations at a deep enough level, and given sufficient knowledge about potential SOS partners, value elicitation can begin.

In Value-Based Engineering and in IEEE 7000™, this value elicitation is guided by three grand ethical frameworks that have shaped European thinking over the past 2,500 years: virtue ethics, duty ethics and utilitarianism.

Virtue ethics

Virtue ethics is probably the oldest and soundest philosophy of ethics. First laid out by Aristotle in his Nicomachean Ethics (Aristotle, 2000), it guided intellectuals' thinking and people's striving for *eudaimonia* for almost 2,000 years, until the time we now call "enlightenment." Eudaimonia, in essence, means to live a "good" life, with the ultimate goal of reaching what Abraham Maslow would call "self-actualization" (Maslow, 1970). But self-actualization with a view to eudaimonia is inherently social and not individualistic. Aristotle believed that reaching eudaimonia in life could only be achieved through a mature and wise character – a character that cultivates virtues and benefits the community, displaying qualities such as prudence, justice, courage, moderation, generosity, gentleness, kindness, honesty, etc.

Because virtues are bound to individual behavior, they can be defined as the "positive value of human conduct" (p. 24 in IEEE, 2021a). In essence, all virtuous behavior is "golden-mean" behavior; there is no striving for excess. For example, when Aristotle described generosity, he positioned it between the extremes of wastefulness and greed. When he described courage, he positioned it between foolhardiness and cowardice. Foolhardiness and cowardice are examples of vices that are wise to avoid (for an overview of Aristotelian virtues and golden mean behavior, see page 43 in Spiekermann, 2016 as well as Aristotle, 2000).

Aristotle's writing about good and bad behavior or character influenced artists and thinkers of the grand world religions of Judaism, Christianity and Islam. An artistic example is Giotto's illustration of virtues and vices painted on to the walls of Padua's Scrovegni chapel (built in the fourteenth century) (Figure 5.3 is based on some of these pieces of art). But looking into the spiritual traditions of non-Western regions, it is remarkable how many of the core ideas around what it means to be wise and good in character overlap with the Aristotelian thinking (Vallor, 2016).

Intercultural comparative philosophy has shown, for instance, how concrete virtue ethical concepts recur in the grand spiritual traditions of the East (Vallor, 2016): A virtuous person, while called *phronimoi* in the Aristotelian tradition, is called a *junzi* in Confucian ethics. In Buddhism, a person of exemplary virtue is called a

88 — Chapter 5 Value-Based Engineering phase 2: value exploration

Figure 5.3: Virtues and vices adapted from Giotto (from top left to right: 5.3 wisdom, 5.3 moderation, bottom left to right: 5.3 envy, 5.3 fickleness).

bodhisattva – a person who seeks (and is close to) enlightenment (Figure 5.4). The Japanese know the "bushidō code," which every samurai (warrior) has to follow. It includes the seven virtues of integrity, respect, courage, honor, compassion, sincerity and loyalty (Figure 5.5).

Compare this bushido code to the list of virtues compiled by Benjamin Franklin (1706–1790), a US-American politician, inventor and thinker, who noted his striving for the 13 virtues of temperance, silence, order, resolution, frugality, industry, sincerity, justice, moderation, cleanliness, chastity, tranquility and humility. There is some overlap between the Japanese and American list of qualities that make a good person, especially in such values as sincerity and humility. However, in line with Franklin being the child of an individualistic culture, his list is on the whole much more self-centered than community-centered.

Figure 5.4: Bodhisattva.

What unites all these virtue-ethical traditions is that they see humankind's progress in people's self-development –a development that leads them to form habits of moral excellence as well as to practical and moral wisdom. But what constitutes moral excellence? And when is a decision wise? Since life is so varied and each moment actualizes non-identical repetitions of events, virtue ethics builds strongly on the ability of a person to adapt in her unique way to situations and be able to deal with actualities as they ought to be interpreted. Scheler expressed precisely this idea, writing: "The person is a continuous actuality. The person experiences virtue in the mode of the 'being-able-to' of this actuality in regard to something that 'ought' to be done"

(p. 85 in Scheler, 1921 (1973)). In other words, virtue ethics trusts in human capability and ability to grow into virtuousness. And the virtues themselves are the names given by the respective cultures to describe this ability shown, in myriad forms of excellent behaviors. This emphasis on people effectively distinguishes virtue ethics from the two other ethical traditions – utilitarianism and duty ethics – that are more rule-focused rather than person-trusting.

Figure 5.5: The seven virtues of a Japanese samurai.

Why are virtue effects so relevant for technology design? Technology design has an influence on human habits and character formation. As Don Ihde and Lambros Malafouris attest: "Materiality and the forms of technical mediation that humans make and use are not passive or neutral but actively shape what we are in a given historical moment" (Ihde & Malafouris, 2019). As mentioned in the introduction, Facebook – in the way it was initially designed – has bred envy (Krasnova et al., 2015) as well as hate (Munn, 2020). Psychiatrists warn that whole generations might develop the vice of inflatedness, or what Elias Aboujaoude calls "e-personality" (Aboujaoude, 2012). In line with this, former employees of Silicon Valley tech-companies have voiced the concern that the current IT systems have the potential to "degrade humanity." It is against this background that VBE and IEEE 7000™ ask:

> **What are the negative implications of the system for the character and/or personality of direct and indirect stakeholders – that is, which virtue harms or vices could result if the system was implemented at scale?**[48]

Utilitarianism

Virtue ethics focuses on the impact of technology on people. But an SOI may also exercise an influence on other more abstract stakeholders, such as nature, society at large, a town, etc. There may be economic consequences resulting from a system that are of importance to the organization wanting to deploy it. Therefore, person-centric virtue ethics alone are not sufficient to get a full grasp of the ethical implications of a system. For this reason, VBE embraces utilitarianism as the second framework for eliciting the value effects of a technology. It does so in line with the general technology assessment literature (which has been criticized for too often relying exclusively on utilitarianism) (Grunwald 2017, 140).

Utilitarianism is a strand of moral philosophy that originated in eighteenth-century England with two philosophers, Jeremy Bentham (1748–1832) and John Stuart Mill (1806–1873) (Figure 5.6). They argued that decision makers should consider the consequences of their decisions by weighing, in almost mathematical terms, the positive and negative outcomes these entail. These outcomes can be added up and weighed on an individual level by asking the question: "What effect(s) will *my* doing this act in this situation have on the general balance of good and bad?" (p. 34 in Frankena, 1973). The goal of this outcome equation is to maximize happiness. Mill himself famously said: "Everyone ought to act so as to bring about the greatest amount of happiness for the greatest number of people."

But such an individual-level questioning of everyone regarding possible consequences, which is also called "act utilitarianism," might be difficult to apply to a technology context, where a whole organization is involved in a flow of design and development activities, resulting in a system that may influence entire societies. Therefore, VBE and IEEE 7000™ use general utilitarianism to analyze the consequences of a system, asking:

Everyone ought to act so as to bring about the greatest amount of happiness for the greatest number of people (John Stuart Mill)

Figure 5.6: John Stuart Mill (1806–1873).

What are all the thinkable positive and negative consequences you can envision from the system's use for direct and indirect stakeholders, if the system was implemented at scale?[49]

In applying utilitarianism, it is important to note that consequences, which Mill coined "utils," were not conceived to stand for monetary utility only. This is what the economic literature developing in subsequent centuries largely made out of the philosophical theory. Instead, Mill and Bentham embraced any form of "pleasures" and "pains," which they advised to rank (or weigh) according to their intensity, duration and certainty. A decision is good if it maximizes pleasures while minimizing pains for the greatest number of people.

Research shows that using utilitarianism for a system's ethical assessment opens stakeholders' minds to considering the greatest number of values that could potentially be impacted by a system (Bednar & Spiekermann, 2022). That said, it is important to also be aware of the criticism that utilitarianism as an approach has received from philosophers such as Alasdair MacIntyre (MacIntyre, 1984) (Figure 5.7). MacIntyre argued that the choice and weighing of consequences for a utilitarian decision can be arbitrary – that is, an expression merely of personal preferences and attitudes – and thereby void of any standard for the good. Happiness, he argues, is not a standard for the good. It does not allow us, for instance, to judge in a situation whether justice is more important than freedom. Furthermore, bad actions can be justified on the back of good consequences (for example, lying). Consequently, MacIntyre has reproached utilitarianism for being a "pseudo-concept" that enters an element of arbitrariness into our moral culture and disguises the true challenges of ethical decisions behind

"Which pleasure, which happiness ought to guide me?"
(Alasdair MacIntyre)

Figure 5.7: Alasdair MacIntyre.

a simplistic rule. Utilitarianism would, for instance, allow for simply weighing corporate profit made from a system as more important than the privacy or security of the people using it; or it would allow taking any lofty principle, such as the "progressiveness" or "productivity" of an organization, as an argument to justify an investment into a questionable technology. The latter issue has been criticized by Thomas Nagel, who noted that utilitarianism can promote such lofty principles, which, he argued, are expressing a "view from nowhere" (Nagel, 1992).

VBE scholars should be aware of these shortcomings of utilitarianism and should therefore never use it as the single source for understanding the spectrum of positive and negative values implicit in a system. Instead, the virtue ethical perspective described above ensures that a "view from nowhere" is balanced and that those individual character implications of a system, which are often neglected by general utilitarian calculus, are recognized. Moreover, in a duty-ethical analysis, VBE teases out those value principles that are likely to have a universal moral claim instead of just being the result of an abstract happiness calculus.

Duty ethics

Duty ethics is a branch of moral philosophy originating in the writings of Immanuel Kant, an eighteenth-century German thinker. Kant (1724–1804) is regarded as one of the most influential thinkers of enlightenment (Figure 5.8). He wanted to create a universal justification for moral actions. In order for moral decisions to be rational, he argued (similar to MacIntyre) that the consequences of an act are subject to volatile

"Act only in accordance with that maxim through which you can at the same time will that it become a universal law."
(Immanuel Kant)

Figure 5.8: Immanuel Kant (1724–1804).

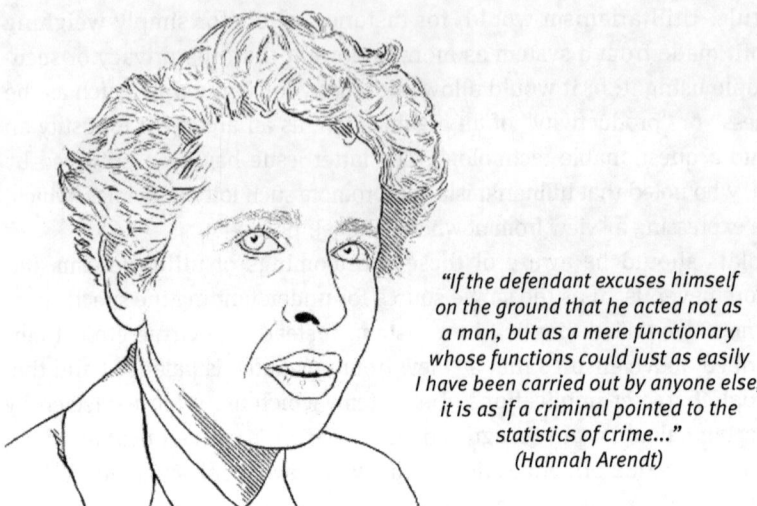

"If the defendant excuses himself on the ground that he acted not as a man, but as a mere functionary whose functions could just as easily I have been carried out by anyone else, it is as if a criminal pointed to the statistics of crime..."
(Hannah Arendt)

Figure 5.9: Hannah Arendt (1906–1975).

ideas of human happiness and cannot serve as a reliable moral compass. Instead, a moral obligation, which he called a "categorical imperative," can be justified only by something that is a universal principle in itself. So Kant formulated one categorical imperative that all more specific actions should conform to: "Act only in accordance with that maxim through which you can at the same time will that it become a universal law" ((Kant, 1785/1999), 73, 4:421).

Note the use of the word "maxim" here. For Kant, maxims are not just any value. Maxims are the highest personal principles governing one's own life; personal principles that one wishes for and acts upon. However, these principles are not chosen arbitrarily, but should be of such fundamental importance that a person would want them to become a universal law. An example is the value maxim of honesty that one might find personally important, but that crucially also has some universal merit.

Kant's categorical imperative also states that the duty to perform moral action arises out of the universal law of respect for other people. If one does not lie because one does not want to get caught, then the act is not morally worthy. Why is the motivation "to not get caught" not morally worthy in Kant's eyes? This kind of behavior is not morally worthy because the motivation for action is not the right one. The motivation for action, according to Kant, must come from respect for human dignity. Thus, building a computer system in a transparent and safe way only because one has to by law, is not of moral worth. The motivation must be genuine.

Interestingly though, Kant's argumentation can be twisted. Take the example of the decision to invest in a fully automated self-check-in system at an airport that is to replace human-staffed check-in desks. An executive could argue that the work of the ground staff at airline check-in counters is boring and demeaning and thereby justify a lay-off decision with the categorical imperative. The executive's argumentation would, in this case, show respect for the dignity of his employees and their potential desire to get more work fulfillment elsewhere. But it overlooks that in the name of this "universal good," employees pay the price of unemployment, insecurity and distress. What is more, corporate executives can be so deeply convinced that profit is the highest maxim a company exists for that they see their sole role as functionaries in the service of monetary gain. They thereby forget about the many other duties they have vis-à-vis employees and communities where they operate. A superficial use of Kant's categorical imperative hence can lend itself to a subjective and estranged interpretation of what is good for humanity. There is a threat that evil is legitimized when supposedly universal values that may, in truth, be problematic, are heralded as good (MacIntyre, 1984). A dramatic historic example of this unfolding was portrayed by Hannah Arendt in her account of Adolf Eichmann's trial in Jerusalem. Being on trial, Eichmann – who had organized the logistics for the transportation of Jews to concentration camps – argued that he had acted in line with Kant, believing that Hitler's "maxim" of "Aryanism" would be of universal value (Arendt, 1965 (2006)) (Figure 5.9). We know, of course, that this is simply not the case. Arendt drew the critical conclusion that immoral behavior cannot be justified with the functional role one has or holds when conducting it. One's own moral conscience should continue to question what is good or bad.

The potential abuse of the categorical imperative is one of the reasons why twentieth-century scholars have suggested that it only works as long as it is embedded in a virtuous culture that shares an uncontestably good set of values (MacIntyre, 1984). Some scholars like William David Ross (1877–1971) set out to compile these. He believed that good human relationships only rely on a few behavioral duties (Skelton, 2012):
- fidelity (the duty to keep our promises);
- reparation (the duty to act to right a previous wrong);
- gratitude (the duty to return services to those from whom we have, in the past, accepted benefits);
- the duty to promote a maximum of aggregate good; and
- the duty of non-maleficence (the duty to not harm others).

VBE emphasizes that technology decisions require few maxims, even though these are not pre-specified as Ross suggests. With a view to a concrete SOI, stakeholders are asked:

Which values and virtues would <u>you</u> consider as so important in terms of <u>your</u> personal maxims that <u>you</u> would want their protection to be recognized as a universal law and that should therefore be respected by the system of interest; in particular, if this system was implemented at scale?[50]

In line with Arendt, innovation teams should look at their (your!) personal maxims here, because duty ethical reflections can easily lead stakeholders to repeat other peoples' or societies' norms, which seem to be set as universal norms (just as Aryanism was a promoted norm in Nazi Germany). For example, corporate stakeholders might think that profit, or growth, or labor productivity should be a maxim for a company. But this is not what is being asked for here; one's personal values of highest importance – and not simply the values of the prevailing economic establishment – are what is demanded.

Respecting diverse world cultures and ethics

Every world region in which a system is going to be deployed might have stakeholders describe maxims that are reflective of their own regional culture. For example, the values of conviviality and community might be considered as a maxim more often in collectivist cultures than in individualistic cultures. As a result, VBE is sensitive to values prevalent in a region and does not follow one cultural template. It embraces the fact that stakeholders from the Eastern or Southern regions will point to maxims that differ from those avowed by Northern or Western stakeholders.

Looking at the three ethical frameworks – virtue ethics, utilitarianism and duty ethics – it becomes clear that in applying them, regionally different value expectations on a system will be collected, reflecting the respective cultures. That said, criticism might still be voiced over the three philosophical frameworks originating on the European continent. Perhaps, other regions in the world have a completely different approach to finding the right path to good behavior and good systems. In order to embrace this possibility, IEEE 7000^{TM} invites project teams to ask whether a fourth philosophical, spiritual or religious framework might additionally be capable of identifying the region's respective value spectrum. Cultural traditions have their unique ways of framing ethics, and may thereby shed further light on what matters to stakeholders. These should be used in projects where this is deemed important.

Finally, it should be noted that the order in which the three ethical theories are applied when using VBE, diverges from the order chosen in this chapter. According to IEEE 7000^{TM}, utilitarian reasoning should be pursued first, because it returns the greatest number of values, thereby giving the freest rein to creative thinking. This expansive reflection is followed by the refinement and extension of the value list with virtues that tend to be neglected by utilitarian reflections (Bednar & Spiekermann, 2022). In the third and final step, duty ethics is added, which leads stakeholders to focus on the highest intrinsic values that they would find to be most important and universally expected from a system. The earlier two analyses prepare the ground for this duty-ethical reflection and concentration.

Practical issues in value elicitation

Value elicitation can tangibly happen in different ways. The project team, participating in a workshop along with the stakeholder representatives and the Value Lead, could write down their thoughts on post-its and freely exchange views on the ethical dimensions of the project. Alternatively, they might work individually in a highly structured manner via a table format, such as the kind depicted in Figure 5.10. Potentially, both approaches can also be combined. In the following, it is assumed that the table structure is used, because this is the method most convenient to create transparency around a VBE effort. The table contains columns to describe an SOI's value effects on different stakeholders and jots down ideas for changing the SOI's concept of operations.

Utilitarian Analysis (TM Example)				
# Stakeholder affected?	Description of Value Effect	Value Name	harmed? fostered?	Change to ConOps?
1 patient	A patient who feels really weak does not need to walk to the doctor where he might even infect others in the waiting room	Comfort	fostered	–
2 patient	A patient who feels really weak does not need to walk to the doctor where he might even infect others in the waiting room	Health	fostered	–
3 student patient	Students who are not sick, but need a sick leave for their exam day, might have a way to remotely fool TM doctors and easily get the sick-notification	Honesty	harmed	request in the conversation guide with students a confirmation of no exam conflicts
4 ...	...	...	...	...

Figure 5.10: Value elicitation matrix.

Virtue Ethical Analysis

#	Stakeholder affected?	Description of Value Effect	Virtue/Vice name	harmed? fostered?	Change to ConOps?
1	specialized doctor practitioners	TM running at scale, practitioners will need to compete to be recommended through the platform	Competi-tiveness	fostered	avoid anonymous recommendations create an open participation process, etc.
#	...	...			...

Duty Ethical Analysis

#	Stakeholder affected?	Description of Value Effect	Personal Maxim	harmed? fostered?	Change to ConOps?
1	specialized doctor practitioners	Specialized doctor practitioners might feel degraded by negative rankings in the TM recommender platform	Dignity	harmed	avoid ranking and bench-marking of doctors, allow for changes in position at regular intervals, etc.
#	...	...	...	...	...

Figure 5.10 (continued)

Naming values

One of the biggest challenges in the value elicitation practice is to nail down the value essence from the various ideas arising in stakeholder discussions. Despite the use of the three guiding ethical questions, project team members and stakeholders are not always able to perfectly distill their thoughts. For example, one stakeholder reflecting on the TM case wondered: "What if patients are abusing the video chat with doctors and lie about their true health condition just to get a sick-note? Doesn't TM's virtual encounter encourage such a lack of accountability?" Looking at this statement, only one relevant value is directly named: accountability. The issue of honesty, however, was hidden in the description and it needed to be teased out. This is typically the task of the Value Lead.

Using a table structure as depicted in Table 5.10 can help the Value Leads' work. The table structure invites the description of each positive or negative value effect separately. In the same row as this description, it is noted which stakeholder(s) would be affected. Most importantly, each description of an ethical theme (arising from the SOI) is captured in the same row by assigning a name to the value implied – "calling the child by the name," as it were. This act of naming a value with a noun is important for several reasons. One is that participants are encouraged to focus their thinking as much as they can and be precise about what they intend to express. The other is that the name chosen by the stakeholders and the project team is a core anchor for the Value Lead to later recapitulate what the participants really meant to say. Moreover, the table invites participants to complement their (especially negative) value descriptions with suggestions on how to constructively improve the SOI's concept of operations. This allows for attaining SOI improvement alongside the value elicitation process. Often, such suggestions for improvement support the Value Lead in her effort to understand stakeholders' concerns. They mirror the stakeholder's concern from an additional angle.

Capturing the value space without bias

One of the biggest challenges for a Value Lead's work is to not project her or his own views and potential criticisms onto the material collected, which might well distort the "truth" of what was really said by the stakeholders. To speak with the tongue of the monk Evagrius Ponticus (345–399), the Value Lead needs to keep an inner stance of "apatheia"; that is, to not feel tempted to be drawn to one side (or certain arguments) from the very start, but to listen and try to truly understand what the stakeholders say.[51]

A practical way to reduce the risk of distorting stakeholder perspectives is to ensure that in the table structure, a rule is followed; that is, "one value per row." If a stakeholder statement then touches upon multiple values (which is often the case), then an additional row in the table structure is used. The original statement is copied into that additional row and an additional value name is added to it. This practice prohibits Value Leads from opinion-led choices on which values to select from a statement. Another way to avoid biased intrepretations is to scrutinize the objectivity of the described value effect: Is the value effect described a verfifiable fact, directly observable in the SOI? Or, is the described effect a general fear associated with the SOI in general. If the latter is the case, the door is open for capturing biased value effect phantasies instead of fact-based issues in the SOI. To avoid bias, Value Leads are well advised to stick to concrete SOI feature facts observed. This helps them to then also find fact-based value (quality) names.

When using a table structure like the one depicted in Figure 5.10, Value Leads are well advised to offer the table's template with as many empty rows for value statements as possible. This is because a prior limitation of the number of description rows signals to participants that only a few important ideas are sought from them. Instead, quite the opposite is true: Value elicitation should be as complete as possible. Everyone should try to think about everything that might possibly go wrong and should consider anything that might create positive values.

This latter point, to think about as many positive values created by the system as possible, might come as a surprise to those who think that VBE is primarily there to prohibit harm, just as technology impact assessment methods aim to. But this would be a misunderstanding of VBE, and also of ethics. Ethics is just as much about doing the right thing and pursuing the good, the true and the beautiful as it is about avoiding the bad. Therefore, value elicitation is keen on understanding the values fostered by an SOI just as much as it is interested in being aware of the values potentially being harmed.

There is only one exception to this positive thinking embedded in VBE and IEEE 7000™. This is that the method does ask for any positive *virtue* effects of technology use. The main reason for this is that prior testing of the method showed that those participants of elicitation workshops who do not possess a high degree of technical knowledge and realism tend to project unrealistic (and often science-fiction inspired) expectations onto an SOI. They make up fantastic virtue benefits that technology experts know can hardly be achieved. An example of this kind of overly-positive virtue projection is the idea that AI-enabled teddy bears would be better suited to educating children than human parents. The virtues of socialness, empathy and better education easily associated with AI lead to an exaggeration of what machines can do for us. In order to avoid such distortions in the data collected and be as inclusive as possible to non-technical stakeholder representatives, positive virtue effects are not elicited.

Value clustering

Depending on the number of stakeholder representatives and project members involved in value elicitation, the number of values collected varies greatly. Typically, at least 50 values are collected, both in the form of ideal core values as well as value qualities. This shows that there is an incredibly rich space of values surrounding every single technology, and the more stakeholders and critical eyes that are participating in the analysis, the more is seen.

One reason for seeing a great number of values in elicitation workshops is that value issues keep recurring under different value names. In the TM case, for instance, "privacy" was reflected upon in three different ways: Participants used the word "privacy" in their own statements where they described confidentiality issues or their wish for transparent data use. However, without mentioning the word "privacy," they named the value in terms of 1) a desire to control their health data use, 2) the security of health data, as well as 3) the idea of staying anonymous vis-à-vis doctors. Value Leads need to recognize that personal data control, security and anonymity all function here as three distinct value qualities underlying the same core value of patient privacy. Figure 5.11 depicts this bottom-up conceptualization. Recall that a core value is (p. 17 in IEEE, 2021a) "a value that is identified as central in the context of a system of interest . . . [It] is at the center of a value cluster of instrumental or related values and value [qualities]." And a value quality (called "value demonstrator" in IEEE 7000™) is a "potential manifestation of a core value, which is either instrumental to the core value or undermines it" (p. 23).

Given a Value Lead's responsibility to analyze and structure the collected value material, and to distill core value clusters and value qualities from a rich baseline material, it becomes clear that she or he plays a vital role in VBE projects.

One way to determine what is "core" (and what is not) is to count all values of the same name as given by the stakeholders, and to define those that are most frequently repeated as core. This was how the core values of the TM platform were determined, for which 14 value clusters could be discerned (Spiekermann, Winkler, & Bednar, 2019). It should be noted, though, that the most prevalent values found in this quantitative way can overshadow the more essential ones that might be mentioned less often. Therefore, a Value Lead should always trust his or her own judgment of relevance in defining what is core. That said, due to the considerable influence the Value Lead exercises in this way, he or she must confirm with stakeholders and the project team that the core values extracted are in line with the

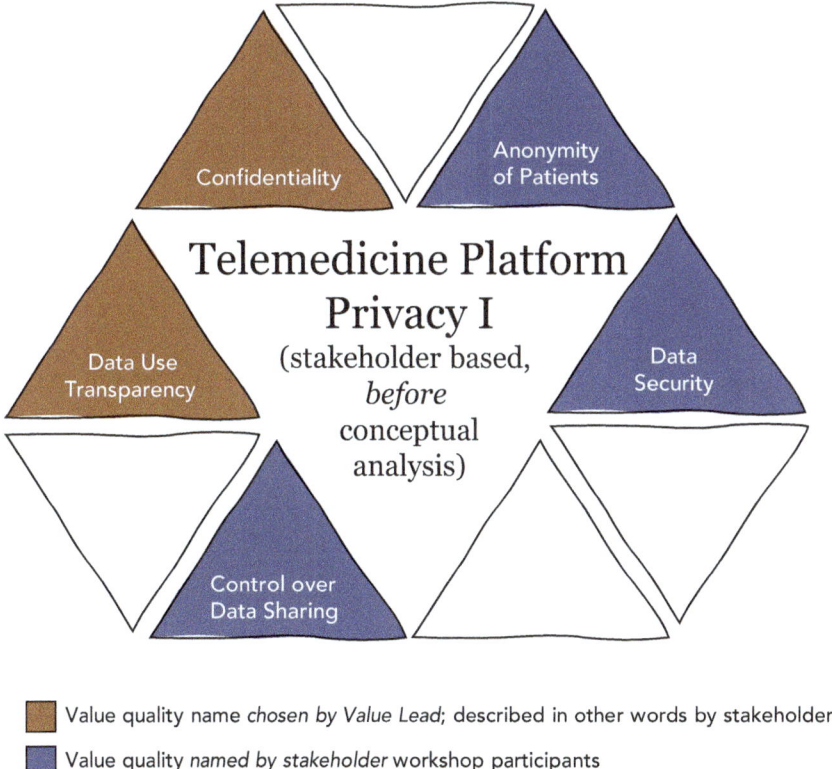

Figure 5.11: The initial core value cluster for privacy of the TM platform.

original value elicitation work in terms of a) completeness, b) relevance and c) jointly understood terminology. The Value Lead must also confirm that the value qualities found to be instrumental to the core values correctly capture the SOI issues raised. In fact, the Value Lead must get an official approval from stakeholder representatives that the clusters and value qualities in them are in line with the stakeholder dialogue. She or he gets this approval through a signature.

Figure 5.12 shows the table structure that a Value Lead can use to rearrange and capture the collected value material in aggregate form and to create a numbered trace capturing the chain of thought. It summarizes all core values in clusters along with their qualities. Note that core values are always positive intrinsic values while value qualities are instrumental to the intrinsic core values, and can therefore either foster or undermine them. In other words, value qualities can be positive or negative, benefiting or undermining the core value. The value trace can be used in the Case for Ethics and in the Value Register IEEE 7000™ that users need to build up in VBE projects. For each core value cluster, there is one separate table with its concomitant qualities. In this way, the 14 core value clusters found for the TM case led to 14 tables of the kind depicted in Figure 5.12.

Core value	benefits from/is harmed by	Value quality	Description of effect	Stakeholder(s)
1. Trust	benefits from	1.1 Reliability of specialist recommendation	Criteria for judging specialist quality	Specialist, doctors recommending
1. Trust	is harmed by	1.2 Lack of Privacy	Lack of security of health records	Patients
1. Trust	is harmed by	1.3 Unjustified exclusion	An unjustified exclusion of specialists	Specialists, TM
1. Trust	is harmed by	1.4 Poor quality	Lack of diagnosis quality due to TM doctors virtuality	Patients
1. Trust	...	1.5 ...	...	...

Figure 5.12: Capture of value qualities related to a core value.

Prioritizing value clusters

Once the core value clusters are identified, project teams need to decide how these should be strategically prioritized for system design. VBE demands the active involvement of corporate leaders in this prioritization because the ranking of an SOI's core values determines the value strategy of a business as a whole – at least where the

SOI is the purpose of an organization's operations. To come back to the garden metaphor (Figure 5.13), what is at stake is the decision of what one wishes to see grow in one's garden and what the garden is there for – to nourish, to heal or to impress. The value strategy built into the SOI determines the purpose of the system, or as what some call, the "value proposition" at the core of a business, product or service.

Take again the TM case as an example. When working with its CEO on value prioritization, he argued that his primary vision for initially founding the company had been to "democratize" medicine: Everyone, he said, should have access to the best doctor specialists, regardless of their social class or financial means. Before founding TM, he observed that privileged patients had better access to the "right" specialist practitioners. Against this background, TM's CEO ranked "patient equality" as the core value with the highest priority for his company. Any patient dialing into his platform, he said, should have equal access to a good healthcare diagnosis and get a referral to a highly recommended (ranked) specialist practitioner. The value qualities consistent with this core value priority were all collected up-front from stakeholders.

Figure 5.13: What is the IT garden there for? What value proposition is sought?.

So, when sorting and ranking value clusters, the CEO understood quite well that an equality strategy would imply the implementation of at least four operational qualities working towards the invisible vanishing point of equality for different stakeholders (Figure 5.14).[52] These included:

Value qualities fostering the core value of equality (green):
1. Patient Inclusion: TM's service would need to be inclusive, allowing poor and uninsured people to use it.
2. Equal Specialist Access: TM's recommended specialist practitioners would need to be ready to treat any patient flowing in from TM, regardless of their insurance or financial status.

Value qualities undermining the core value of equality:
3. Lack of Equity: Patients meeting TM practitioners only through the portal might not have the same quality of care as patients visiting TM practitioners in their real offices.
4. Patient Exclusion: TM would need to ensure that those who do not have a computer or are anxious about using one can still use the service.

A key decision for the CEO was to determine how the value of equality could be balanced with that of privacy, especially since the TM service processes so much personal data (Figure 4.5). What should be prioritized in the subsequent system development?

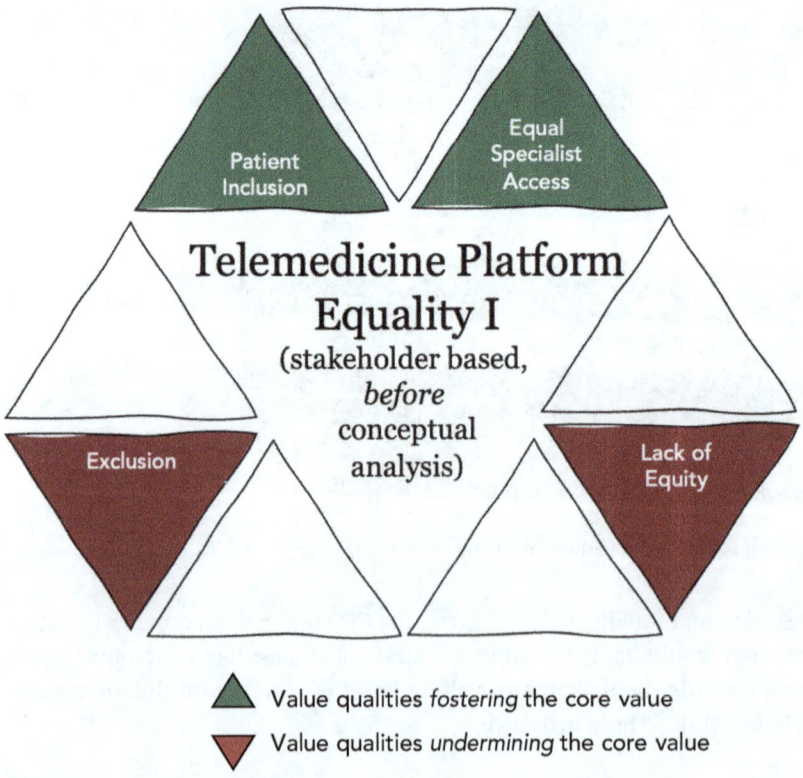

Figure 5.14: The initial core value cluster for equality of the TM platform.

And how would these two fare vis-à-vis the other 12 core values that were part of the ranking exercise, among them being trust, knowledge gain, comfort and reliability of the service (case study). Answering this prioritization question was vital because research has shown that value conflicts in system development teams can only be avoided if value priorities are officially set by top management up-front (Shilton, 2013). If this is not the case, then developers are unsure of what to do first and prioritize in their own work. One engineer might find equality more important than privacy and another, the opposite. Therefore, VBE obliges corporate leaders to set the priorities that everyone can then follow in system development. However, it is typically not easy to decide what comes first: in this case, the business mission priority of equality that the CEO wanted to cater to, the privacy of patients' health data or the other values, like trust, knowledge, comfort and reliability. To answer such a difficult prioritization question, VBE with IEEE 7000™ offers a number of criteria that can help to make this ranking decision.

Criteria for prioritizing core values

The IEEE 7000™ standard lists seven criteria to be considered during value prioritization (p. 41 in IEEE, 2021a), which are all equally important:
I. Stakeholders agree that the SOI is good for society and avoids unnecessary harm.
II. The organization does not use people merely as a means to some end.
III. Organizational leaders can accept responsibility for the value priorities chosen, according to their own personal maxims.
IV. The organization respects its own stated ethical organizational principles, if there are any.
V. The organization can commit to the value priorities in its business mission.
VI. The environment is maximally preserved.
VII. The organization considers existing ethical guidelines.

Inherent in this list of prioritization criteria are four duty dimensions (Figure 5.15): One duty aims for an alignment of the SOI's value priorities with the shareholders' expectations. This is why ranking criterion V says that an organization should be able to commit to the value priorities in the light of its business mission. Let's say an organization committed to "privacy" as its most important value priority. Then, the business mission communicated to shareholders could hardly be to collect as much data (oil) from customers as possible and leverage it for profit. If an organization's business mission was crucially reliant on the commercialization of personal data, it would be hypocritical to prioritize privacy as a core value. If the organization still wanted to rank privacy high in such a setting, its business mission would probably need to be adjusted.

This is exactly what the core value ranking activity achieves: in reflecting on priorities, it is likely that the business mission – as well as strategic priorities – will be changed.

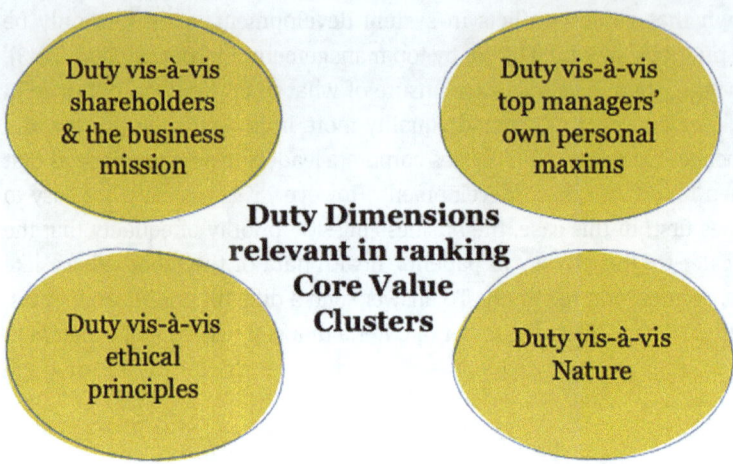

Figure 5.15: Duty dimensions relevant for determining value priorities.

The second duty dimension relevant for prioritizing value clusters is to look at the maxims of the top executives involved in the project (criteria I, II & III). When corporate leaders rank core values, they are asked to reflect on the maxims they personally hold (Kant, 1785/1999). Hence, ranking the values requires them to introspect. It encourages them to only prioritize core values for an SOI that they would *personally* want to become universal in the society. Their *personal* reflection is underscored here because, often, organizational leaders are forced into some role-play on the job. In their role as executives, they often tend to live by values other than those they hold dear in their private lives. Marketing executives were shown, for instance, to be very sensitive about their own private personal data sharing, while justifying this same data sharing when asked on the job as professionals. VBE invites executives to end such ambiguity or dissonance between their private views and their professional attitude. A rank order criterion that helps them achieve this is the presence of other stakeholders who all need to agree that the SOI is good for society and it avoids unnecessary harm (criterion I). Along with these stakeholders, organizational leaders can discuss the joint duty to promote a maximum of aggregate good through the SOI and avoid maleficence (recall the duties described by Ross, 1930). Here, criterion II can also play a role, since it invites organizations to ensure that their value ranking does not use people merely as a means to an end. Automated airport check-in machines may clarify this last argument as an example. Let's suppose two core values were pitched against each other for such a machine. One is the core business value of profitability, achieved through the value qualities of efficiency and cost savings. The other is the core value of customer wellbeing, achieved through the value qualities of customer help and personal address.

In this case, criterion II would have wellbeing ranked above profitability, because efficiency and cost savings are only achieved by lay-offs, effectively regarding people as replaceable – an avoidable means to a business end.[53]

A third duty dimension is that vis-à-vis existing external ethical principles (criteria IV & VII), organizations are invited to compare their preferred SOI core values with potentially existing corporate social responsibility (CSR) principles or other commitments they have made. Examples are IBM's commitment to accountability, explainability and fairness for its technical products or Microsoft's commitment to people empowerment, community and environmental sustainability. Such self-imposed organizational principles show their true worth when they are used in an exercise like value cluster prioritization where they will effectively influence an SOI.[54]

That said, there are not only principles from within an organization that must be considered here for comparison; of greater importance are external laws and regulations. Many technical systems still suffer from their ignorance of regional laws or international human rights agreements. VBE changes this aspect. It ensures that politically established principles are considered in core value prioritization. For example, when an SOI is heavily reliant on personal data, project teams would have the duty to rank the core value of privacy relatively higher in comparison to other values because it has been recognized in so many laws and international agreements. In TM's case, for instance, privacy might be ranked even higher than equality, because breaching it could lead to severe financial penalties that even a great business mission cannot make up.

Last but not least, a rank order criterion is the explicit recognition and respect for nature. If an SOI is recognized to have clear sustainability implications, for instance, due to its power consumption, then this implication for nature would be heavily weighted in comparison to other SOI values.

Resolving value trade-offs

A challenge may arise when despite the given ranking criteria, some core values still contradict each other or are hard to put in a ranking order. They seem to require that project teams make trade-offs. Value trade-offs are widely regarded as a dilemma. Scheler's Material Value Ethics, which is the basis for VBE's value understanding, does not share in this conflict-laden view on values. Instead, it sees values in a natural hierarchy (Scheler, 1921 (1973)). More precisely, Scheler argues that ethical behavior is essentially constituted by continuously choosing and realizing higher values over lower ones. He outlined how
1. . . . the relative endurance,
2. . . . depth and
3. . . . indivisible nature of values are indications for their relative superiority. In addition

4. ... their relative independence from value-bearers and
5. ... their intrinsic, non-instrumental nature
   ... can be indications for their rank (p. 86ff).

Figure 5.16 summarizes the criteria and gives examples on how to interpret them. The table is also contained in IEEE 7000™'s Annex B.

Values are higher ...	Examples
... the more they endure (has nothing to do with absolute time, but with the persistence of a value, the eternality of a value)	Love is higher than enthusiasm Happiness is higher than convenience
... the less they are extensible or divisible	A piece of art cannot be divided, which is why it is of higher value than a piece of bread; beauty as a phenomenon is of higher value than attractive haircut
... the less they are founded through other values (classical distinction between intrinsic and extrinsic values)	Dignity is a higher value than politeness, which caters to dignity
... the deeper the satisfaction connected with feeling them	A deep life satisfaction vs. being happy while being on a walk
... the less the feeling of them is relative to the positioning or existence of a specific bearer of feeling or preferring	... moral values (e.g. truth) are higher than a value such as convenience, that needs a bearer (a situation or thing that is convenient)

Figure 5.16: Max Scheler's criteria for determining the rank/height of values.

Two values from the TM case illustrate how some of these rank-order criteria of Material Value Ethics work: Doctors' collegial respect that manifests itself in mutual analog recommendations today could be traded for the efficiency provided by a recommendation database operated by the TM platform. Efficiency, however, is a lower value than respect. Unlike respect, efficiency has little intrinsic worth. Efficiency is always instrumental to some other value, like profitability. Respect, in contrast, cannot be doubted as worthy in itself. It also unquestionably leads to deeper satisfaction in people than efficiency does. Finally, respect is likely to endure longer than process efficiencies that can easily erode, once minor changes are made to a process. Against this background, the TM platform would be well advised to prioritize doctors' mutual respect as a core value over and above the efficiency gains potentially inherent in a database solution.

Conceptual analysis of core values

After value elicitation and prioritization, VBE requires a refined conceptual analysis of those core values that have been prioritized. Conceptual analysis of the elicited core values is a hermeneutical exercise that looks at the value qualities identified bottom-up by the stakeholders and completes these top-down with value qualities known in the literature and in the law as they are relevant for the context at hand. As value philosopher Ibo van de Poel outlines: "Conceptualization of value is the providing of a definition, analysis or description of a value that clarifies its meaning and often its applicability" (p. 20 in van de Poel, 2018).

For example, the core value of privacy might have been characterized in the TM case bottom-up, as depicted in Figure 5.11. Five value qualities were seen by stakeholders, including patients' control over their health data, data security, the possibility of remaining anonymous as a patient, confidentiality and data use transparency (marked in blue and brown in Fig. 5.17). However, these five value qualities alone are not sufficient for building a privacy-sensitive platform. From an expert perspective,

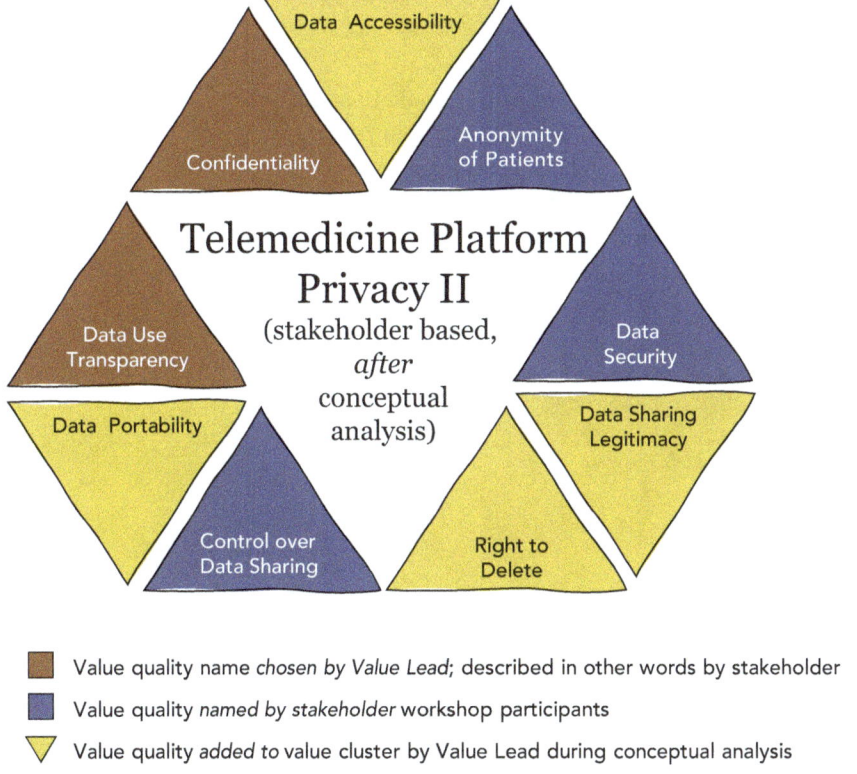

■ Value quality name *chosen by Value Lead*; described in other words by stakeholder
■ Value quality *named by stakeholder* workshop participants
▽ Value quality *added to* value cluster by Value Lead during conceptual analysis

Figure 5.17: The privacy cluster of the TM platform after conceptual analysis.

privacy must additionally be understood in light of the General Data Protection Regulation (GDPR) if the platform were to be operated in Europe. Therefore, various value qualities would need to be added to the cluster, such as data portability, data accessibility or the assurance of legitimacy of any further health data use. Effectively, all principles of the GDPR become value qualities. In this vein, it would also become apparent in conceptual analysis that confidentiality is effectively only one dimension of data security. So, the bottom-up cluster (Figure 5.11) is corrected for this redundancy and incompleteness (Figure 5.17). If external sources on a core value's meaning – such as an applicable law – are available, they provide project teams with a nuanced view of what a value means and what additional factors ought to be considered during system design later. Figure 5.17 illustrates this for the privacy value example of the TM platform.

The final value clusters, prioritized and refined in their value quality structure, constitute an overall ethical value strategy, which should be formulated in an Enterprise Ethical Policy Statement that is signed by both top executives and the stakeholders involved.

Check Questions

- What are personal maxims and what should they not be?
- What does VBE do to accommodate different cultures?
- Would Benjamin Franklin be a perfect Samurai?
- Why is utilitarianism not enough to understand the value spectrum of an SOI?
- What discerns higher from lower values?
- What tasks does a Value Lead have and what are the challenges of her/his work?

Chapter 6
Value-Based Engineering phase 3: ethically aligned design

The last block of VBE work that leads to an Ethically Aligned Design starts with the formulation of the so-called "Ethical Value Requirements" or "EVRs." An EVR is defined in IEEE 7000™ (p. 18) as an "organizational or technical requirement catering to values that stakeholders and conceptual value analysis identified as relevant for the SOI." Thus, an EVR is, in its nature, more concrete than a value quality (Figure 6.1). It is the "bridge" between the values relevant for the SOI and the concrete, specific system-level requirements that will guide the enabling of these in the SOI.

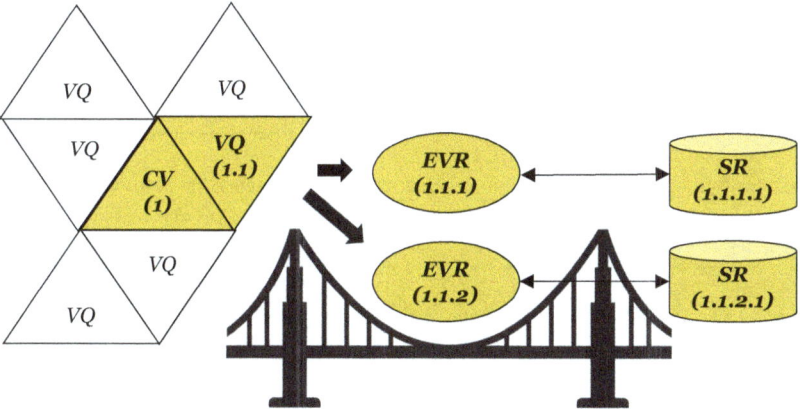

CV = Core Value; VQ = Value Quality; EVR = Ethical Value Requirement; SR = System Requirement

Figure 6.1: EVRs serve as a bridge between value qualities and system requirements.

Ethical Value Requirements

What EVRs are can be illustrated by taking the value quality example of "informed consent" in the TM case. An informed user consent is needed for any privacy-friendly system. EVRs associated with informed consent include: (1) meaningfully and comprehensively describing personal data processing activities to TM users, (2) truly and voluntarily obtaining patients' processing consent, (3) offering patients easily accessible options to decline consent, etc. Looking at this list it becomes clear that EVRs are actually organizational, technical or social *measures* that should be taken in order to protect or foster a value quality. These measures address the risk of a positive value quality not unfolding as desired, or a negative value quality taking

https://doi.org/10.1515/9783110793383-006

effect. Therefore, EVRs have also been called "risk treatment options" in IEEE 7000™ (p. 45 in (IEEE, 2021a).

Figure 6.2 contains an exemplary list of EVRs that were found for TM's prioritized core value of giving patients a fair and equal access to medical care (equality). The table illustrates how the chain, from the core value of equality to value qualities and to EVRs, can be documented through a numbered trace by organizations. This transparent documentation should be included into a reporting tool called the "Value Register" (p. 23 in IEEE, 2021a).[55]

EVRs provide direction regarding what the system-level requirements should cater to. For example, EVR 1.3.1 states that TM should not insist up-front on health insurance credentials for medical service access if it wants to cater to its social equality strategy. The technical system-level requirement that is later derived from this EVR could then ask developers to have the platform's interface only ask for insurance credentials once online consultation is completed, and not as part of the login process.

Core value	fostered/ harmed by	Value quality	Ethical Value Requirements (EVRs)
1. Equality	fostered by	1.1 Equal Specialist Accessibility	1.1.1 Only include recommended specialist partitioners who are at the same time willing to treat at least five patients without insurance per week
1. Equality	harmed by	1.2 Patient Exclusion	1.2.1 Include an affordable telephone call-based channel in TM's diagnosis service
1. Equality	fostered by	1.3 Patient Inclusion	1.3.1 Do not insist on health insurance credentials up-front for digital health service access
1. Equality	fostered by	1.3 Patient Inclusion	1.3.2 Accept diverse forms of payments; also other than those which require a bank account
1. Equality	...	1.4 ...	...

Figure 6.2: Ethical Value Requirements (EVRs) and their value origins.

Formulating EVRs and their criteria

When EVRs are put down in a Value Register, each of them should be concretized and qualified. Examples would be the affordability of TM's call-based channel or the criterion that only those specialists should be included who are willing to treat at least five uninsured patients (see Figure 6.2). In other words, EVRs are formulated as concrete testable thresholds, assumptions or constraints. Remember that a core value is accessible to human perception only through its value qualities. But to enable these value qualities, EVRs need to fulfill stakeholder expectations and reach their threshold levels for value fulfillment. IEEE 7000™ recognizes this need by asking organizations "to identify and record assumptions and constraints" in the Value Register that are associated with EVRs (9.3.a.2 & 4), p. 45 in IEEE, 2021a). Figure 6.2 shows how this can be done.

One might wonder why it is so important to record EVR assumptions and constraints. Why does IEEE 7000™ emphasize this through an extra mandatory task? The answer to this question becomes clear when looking at TM's privacy value again. As outlined above, an important legal value quality for privacy is that people sharing their personal data are consenting to its use in an informed way. So, an EVR for the privacy quality of "informed consent" in TM's Value Register would ideally read something like this:

> Ensure that patients can give consent to their health data processing in an easy and informed way, whereby 'informed' means that the information provided is instantly accessible and comprehensible for laypeople.

The recorded EVR constraints are that policies are easily accessible and comprehensible to laypeople. Compare this to what digital players are doing today: they "inform" their customers of personal data collection through lengthy, non-accessible and complicated "uncontracts" that only legal experts can understand (Zuboff, 2018). The example illustrates that without adequate EVR assumptions and threshold levels built into EVR formulation, IEEE 7000™ users could easily get around ethical conduct: They could tick the box of having identified an EVR and done something in their organizational set-up to address it, such as getting patient consent; however, the way they got it would not be in line with the ethical expectations of stakeholders.

That the EVR's assumptions and constraints – which one might also call ethical threshold expectations – are in line with ethical expectations is ensured by another IEEE 7000™ activity that requires stakeholders to validate them. The partially external and critical stakeholder representation must ensure that EVRs as well as their constraints and assumptions are wisely chosen and detailed (9.3. b), p. 45 in IEEE, 2021a).[56]

Finally, it should be noted that many IT services might need to run through certification procedures prior to being allowed access to a market. For instance, AI services' fairness and non-discrimination, or its degree of transparency, might be tested. EVRs' constraints and assumptions are those indicators against which such certification testing can be done.

Multiple paths of ethically aligned design

Once EVRs are identified, documented and validated with their assumptions and constraints, the corresponding technical and organizational system requirements can be identified and documented.

Recall that "system" design in IEEE 7000™ and VBE means "socio-technical" system design. System requirements in VBE are therefore not only technical, but also socio-technical requirements, including, in large part, organizational management measures.[57] People, policies and management can handle many EVRs with little effort at the organizational managerial level without any technical disposition ever being built into the SOI. Take the EVR 1.1.1 for the value of equality in Figure 6.2. It tells TM to "only include recommended specialist practitioners who are at the same time willing to treat at least five patients per week, without insurance." This is an EVR that does not require any technical system feature. Instead, the EVR calls for the set-up of an organizational process that is responsible for seeking philanthropic specialist practitioners, monitoring their enduring willingness to treat the poor and deleting those specialists from TM's database and network who turn down poor TM patients for treatment. Such an EVR can be immediately handled by TM's management without any further analysis. For this reason, the first "Ethically Aligned Design" path on the bottom left of Figure 6.3 is titled "immediate organizational design." Note that this organizational design path is not explicitly included in IEEE 7000™.

Where organizational management action does not suffice, VBE offers two alternative risk-based technical design processes. This is shown at the bottom of Figure 6.3: What is called a "Standard Risk-based Design Part" (in line with IEEE 7000™ Section 9) and an "Impact Assessment-led System Design Part," as many high-risk systems require one. The former likely applies to most cases in system design. The latter comes into play, however, only when particularly critical values are at stake (such as health, safety or security), when particularly risky systems are built or when the regulator mandates an impact assessment-led system design (as might be the case for some AI systems).

Standard risk-based technical design

The goal of standard risk-based design is that all EVRs that are not able to be handled through administrative measures find their way into the technical roadmap of the SOI. This roadmap is typically maintained by those system and software engineers who are responsible for getting the SOI up and running. It contains all system-level features and capabilities (including requirements) that are scheduled for development.

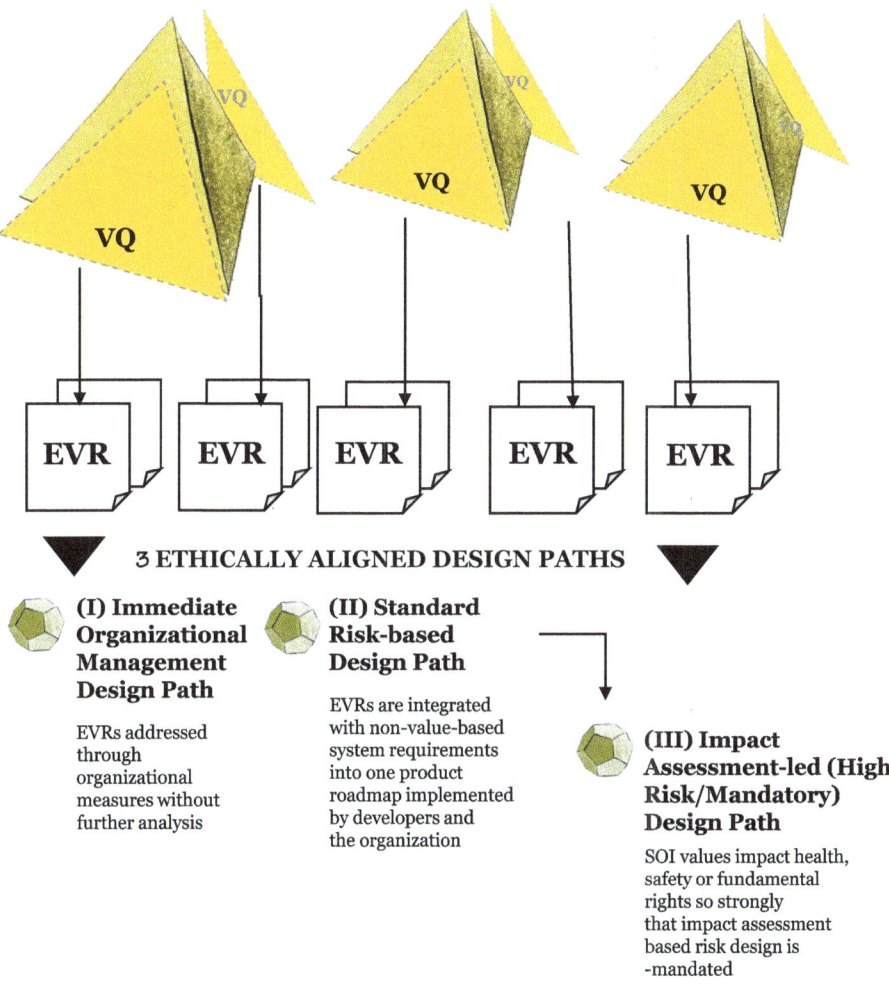

Figure 6.3: Overview of three alternative design paths from value qualities to system requirements.

Initially, technical engineering units in an organization will probably continue to work separately from an IEEE 7000™/VBE effort. After all, a big effort is required to get a system up and running in the first place. However, as outlined earlier, they are required to send a technical stakeholder into the project team, who will be responsible for VBE. Once EVRs are formulated and confirmed by the stakeholder representatives, the time is ripe for the technical VBE shepherd to bring EVRs to the technical units and ensure that they are respected. She or he needs to translate each EVR into the technical system elements and architecture requirements that are not only fit for inclusion in the existing roadmap but also specify the policies and other socio-technical measures necessary to ensure that the SOI will live up to

EVRs.[58] Figure 6.4 illustrates this step, where the two domains – let's call them, "value-based" and "non-value-based" domains – merge.[59]

Conflicts between value-based and non-value-based engineering may arise when the two domains integrate their expectations for the system. These can be minimized, however, if the VBE effort starts early enough, so that non-value-based technical work is neither too advanced nor fixed to specific solutions.[60]

VBE's simple risk logic

Figure 6.4 illustrates that in order to integrate the value- and non-value-based domains, EVRs first need to be translated into socio-technical *system requirements*. This is done via a risk-based approach as many technical computer ethicists have recommended (Gotterbarn & Rogerson, 2005; Spiekermann, 2016). The reason for the risk-based approach is that by conceiving of values as "being-at-risk," development teams are put into a spirit of care and awareness that each value and its associated EVRs might be breached.

What a simple risk-based design approach does is that it asks how each EVR could be put at risk of not being fulfilled. Figure 6.5 illustrates this activity. The risk of an EVR's non-fulfillment comes from concrete "threats." Technical system requirements, so-called "controls," are then developed to address each threat in such

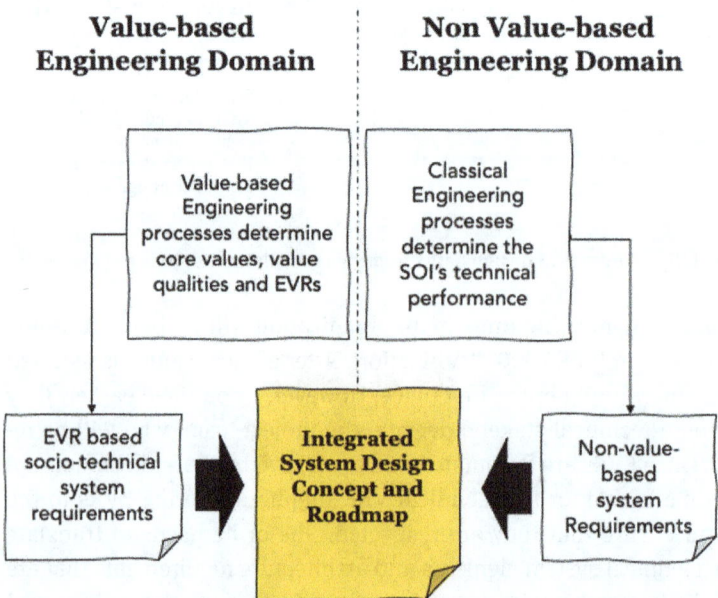

Figure 6.4: Integrating the value-based and non-value-based requirements into one system roadmap.

a way that the EVR is fulfilled. In this way, potential value threats are technically mitigated before they happen.

I. Core value	II. Value quality	III. Ethical Value Requirements (EVRs)	IV. EVR Threat	V. EVR Control (System Requirement)
1. Privacy	1.1. Informed Consent	1.1.1 Meaningfully and comprehensively describe personal data processing activities to TM users	T 1.1.1.1 The privacy policy is not meaningful, disguising information that is really relevant for the user from a privacy perspective (like data sharing practices, external partners and data recipients)	C 1.1.1.1 Offer a complete and honest overview of personal data sharing and processing activities with a view to those areas relevant for patients' privacy perception and judgment
1. Privacy	1.1. Informed Consent	1.1.1 Meaningfully and comprehensively describe personal data processing activities to TM users	T 1.1.1.2 The privacy policy is not easily processible or readable by users; especially not for laypeople	C 1.1.1.2 Offer short layered privacy policy, potentially using (ideally standardized) icons tested for their intelligibility
1. Privacy	1.1. Informed Consent	1.1.2 Truly and voluntarily obtain patients' processing consent	T 1.1.2.1 Users are nudged to press the wrong button that makes them consent to data processing they actually do not want	C 1.1.2.1 Consent interface needs to be designed nudge-free; make any kind of consent option equally easily accessible as a viable choice
1. Privacy	1.1. Informed Consent	1.1.3 Offer users easily accessible and fair option to decline consent	T.1.1.3.1 Users are forced into a coupling deal, such as not being allowed to access the service without consenting to extensive data sharing	C 1.1.3.1 Avoid coupling of service use with data sharing necessity where possible by giving option to only share information necessary for processing

Figure 6.5: Tracing standard risk-based design in table form.

The process of matching value threats with controls, one by one, forces the system and software engineering teams into a state of strong diligence. For example, it was outlined above that an EVR to ensure informed consent requires it to meaningfully and comprehensively describe personal data processing to TM's users. There is a threat that TM's privacy policy might not be meaningful enough, or that it might disguise relevant information from its users. Very long, unreadable and incomplete terms and conditions are still a well-known phenomenon in the web service world, as of 2023. Therefore, a control (mitigation step) against this threat that developers can then implement is to offer a complete and honest, easily accessible, ideally layered and tested policy. Figure 6.5 shows how this can be done, and provides further examples of how the value quality of informed consent can be fostered in a system.

EVR controls are socio-technical system requirements, foreseen for use by developer teams that should know from the way they are written, what needs to be respected in the SOI. It is recommended that each row in Figure 6.5 contains one precise threat and a matching control. One EVR may see multiple threats. And one threat may have multiple controls. One control may also mitigate multiple threats from diverse EVRs. Furthermore, the numbered trace from value qualities to EVRs is continued. This allows the tracing of all value-based system requirements back to the core values they are supposed to support. The IEEE 7000™ standard recognizes this step as follows: "Record each value-based system requirement with a unique reference number, its traceability to an EVR, its associated risks, and related assumptions and constraints" (see tasks 9.3 c) 4), p. 45 in the standard). Figure 6.5 illustrates the threat-control risk logic and how it can be applied to the IEEE 7000™ EVR construct.

Risk-driven requirements engineering in this form is not new to engineering. The spiral model of system design, for example, which constantly iterates system design versions based on a risk perspective, was already established in the 1980s (Boehm, 1988). Privacy and security impact assessments follow a similar threat-control logic (Oetzel & Spiekermann, 2013).

Risk-based system requirements versus risk management

The kind of threat-control logic used to identify risk-based system requirements is sometimes confused and/or conflated with what is called "risk management." Large organizations often have risk managers accompany projects. As the term "management" signals, the activities performed by such managers consist of the measurement, monitoring and handling of all the things that can go wrong in an IT project. For example: running out of budget, not documenting steps sufficiently, not including stakeholders sufficiently, not having put aside enough time, etc. (for a recent survey of such risks in agile software development, see, for instance, Hammad, Inayat, & Zahid, 2019). All such project management risks can of course also play out in VBE, and organizations seeking compliance with the IEEE 7000™ standard are encouraged to engage risk managers in the supervision of VBE projects, depending on the degree of risk inherent in the project they pursue.[61] But, such management risks are not the same as the "risk-based design" described above. In Value-Based Engineering, we use the "threat-control" analysis – which is a risk logic – to derive system-level requirements. This derivation of system requirements with a risk-logic is not the same as project management risk, which is a separate and potentially complementary effort.[62]

The risk-based design in IEEE 7000™

In IEEE 7000™, the risk-based approach to system requirements engineering for EVRs is embedded in the following tasks: (9.3. c), p. 45 in IEEE, 2021a):
1) Analyze the value demonstrators [called "qualities" in VBE] and risk mitigations in the EVRs to identify potential value dispositions.
2) For each EVR or related EVRs, formulate one or more associated value-based system requirements (functional or non-functional) that realize the EVR within the SOI.
3) Identify qualitative or quantitative measurement targets and acceptance criteria associated with each system requirement.

The "analysis" of value qualities ["value demonstrators"] and their EVRs implies looking at how these values can be threatened or put at risk (column IV in Figure 6.5). "Risk mitigation" then identifies controls for these threats (column V in Figure 6.5). These control requirements describe a "system characteristic that is an enabler or an inhibitor for one or more values" (p. 23 in IEEE, 2021a), or what IEEE 7000™ also calls a "value disposition." In formulating these system requirements (column V), project teams must be careful to respect the threshold levels (assumptions and constraints) captured in the EVRs (column III). They do this by formulating EVR controls (system requirements) in such a way that "qualitative and quantitative measurement targets and assumption criteria" are included (9.3.c)3). For example, row 2 in Figure 6.5 notes that the informed consent of a system user can only be given when the personal data processing activities of the system are "meaningfully and comprehensively" described. A realistic threat exists, though, that this is not achieved where a privacy policy is not easily processable or readable, especially not for laypeople (row 2, column IV). A way to mitigate this threat is by using a "short, layered privacy policy" in the interface, which in the long run includes standardized and tested icons that users can recognize. Such short, layered privacy policies are currently a best-available technique (BAT) commonly accepted among privacy experts to reach the (qualitative) target of readability. Therefore, the standard's third point (9.3.c)3) is fulfilled; that is, to "identify qualitative . . . acceptance criteria . . ."

How do project teams know about "best-available" privacy techniques or "BATs" that might help foster any other value like security, safety or transparency? Here, qualified Value Leads come in. Their education must ensure that they are on top of BATs for various values or are able to develop them for a project at hand, in cooperation with stakeholders. Value Leads can find BATs, for instance, in more fine-grained value-specific standards. For instance, the IEEE 7001™ standard details the value of transparency. The Value Lead can also turn to so-called "pattern catalogs," which list best practices for value design, such as the privacy-pattern catalog developed at Vienna University of Economics and Business[63] or the security-pattern catalog developed by KU Leuven.[64] If neither standards nor pattern-catalogs exist, Value Leads can also turn to expert literature. An example is the report on "Understanding algorithmic decision-making," which describes best practices for algorithmic transparency as of 2019 (Catelluccia & Le Métayer, 2019).

Integrating value requirements into SOI's roadmap

Once system requirements are formulated (column V of Figure 6.5), they should be entered into the SOI's integrated roadmap (Figure 6.4). For this purpose, the standard instructs to "Analyze and harmonize the EVR and value-based system requirements with requirements derived from *non-value* driven means, identifying and rationalizing competing or supportive requirements for the SOI" (9.3.d) 2), p. 45).

What does this look like in practice? Often, non-value-driven requirements outline what it is that needs to be done to get an SOI up and running or functionally in place. For example, the technical roadmap of TM might initially specify a functional to-do, like creating the landing page, user registration, doctor database integration, etc. and the list of these functional to-dos will have an associated time plan. This is what is typically called a "roadmap." Value-based system requirements, derived from EVRs, are now added as additional to-dos to this roadmap, and they add information as to how existing and new functional and non-functional to-dos should be put in place, beyond their mere performance and reliability. Value-based control analysis, as outlined above, specifies what to expect from a functionality (beyond reliability). For example, the technical to-do of programming a "user registration" is complemented by the requirement to have a "privacy policy" and a "user consent" as part of that registration. And these two are again characterized in line with the threat–control analysis described in part in Figure 6.5: The privacy policy should be layered, stand out visually, be readable by laypeople, etc. In other words, value-based system requirements recognize the "measurement targets and acceptance criteria" (9.3.c) 3) p. 45 in IEEE, 2021a), in line with the EVR assumptions and constraints. The resulting system roadmap, which then contains these qualified and integrated requirements, is the basis for a development team to implement the SOI. Figure 6.6 illustrates this flow.

This whole risk-based system requirements work is double-checked by the stakeholder group that accompanies the project. It was outlined above how important external stakeholders are for an ethically aligned design and how they are carefully chosen to become part of a permanent verification board accompanying the project, and how they need to regularly check on the ethical decisions taken. This stakeholder group approves the integrated roadmap, with its system requirements, thereby ensuring that

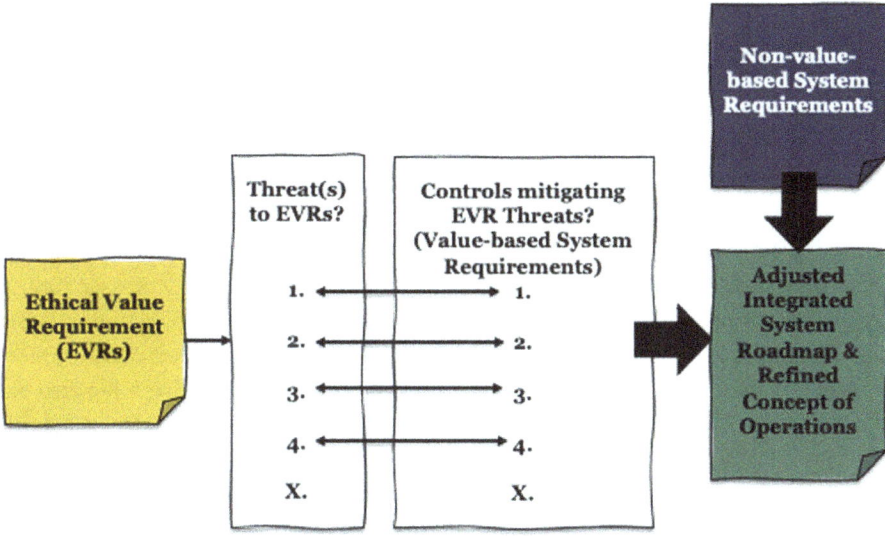

Figure 6.6: A threat-control analysis derives system requirements for an integrated SOI roadmap.

the ethically derived requirements are not deprioritized or swept under the rug. IEEE 7000™ states (see 9.3.e): "Analyze, trace, and record the further handling of value-based requirements in agreement with the project team and stakeholders" (p. 46). The result of this effort must then be validated and recorded in the Value Register, together with a refined concept of operations. The hope of IEEE 7000™ authors was certainly that both the agreement of stakeholders with the overall system roadmap as well as the recording of this agreed decision would encourage organizations to prioritize ethically derived system requirements in a reasonable way.[65]

Ensuring that the whole system is ethical

After value-based and non-value-based system requirements have entered the SOI's roadmap, it is important that the actual design features chosen by a project team are in line with these requirements. For example, there are various privacy patterns that system developers can choose from. Section 10 of IEEE 7000™, therefore, describes activities and tasks that seek to ensure that the whole system is in line with the value mission. Note the importance of the whole system being observed here. Value-based features must be integrated and "harmonized" with the non-value-based features that the technical units have worked on. This can be a political issue when the time schedule or budget for a project's delivery is tight. In VBE projects, everything must be done to ensure that value-based features are not the de-prioritized ones. This is still common practice today due to the Minimum Viable Product guidelines, where developer teams might be at risk of prioritizing operational features to get the system up and running before turning to the "softer" value-based system elements. VBE projects must "harmonize" the VBE-derived requirements with those derived by non-VBE means such that the core values and their value qualities are already represented within the Minimum Viable Product and future iterations of the design (see 10.3)a)2).

Such a harmonization does not only consist in prioritizing development tasks. It also resides in grander decisions, such as the system architecture chosen. For example, privacy-friendly systems might require a more decentralized architecture than would be required from a mere functional perspective. The level of security or user control granted might be higher. Such extra demands on the system's architecture must be brought in line with the functional non-value-based plans for the system design. Stakeholders and the project team should then check once more whether the integrated list of deliverable features does not bear its own risks (IEEE 7000™ 10.3 b)). One such risk that may arise in today's distributed software architectures is a lack of control over externally sourced software elements. Project teams must therefore analyze and specify whether they have sufficient technical and organizational control over their system and the external service elements they couple with (10.3 c)). This control activity seems to be particularly important when an external AI component or a web service is used to provide a functionality with ethical import. It must be ensured that the ethically derived system requirement for it (column V, Figure 6.5) is not compromised. To ensure this, Section 10.3 c) in the standard outlines that each system element (e.g., the AI component) must have an identified owner, must be controlled (10.3 c) 3) & 4), p. 48) and should be tested (simulated or prototyped) for its effectiveness and acceptability (10.3 c) 5). Where there are risks or control issues, an organization must identify and select treatment options (10.3 d), which are then verified and validated with the rest of the design (10.3 e). Note that the validation activity in Section 10.3 e) (p. 48) also includes a monitoring of the design that is then launched onto the market (10.3 e) 8)). Therefore, when system providers or regulators look for a "continuous iterative process of system validation and verification throughout the entire product lifecycle" (EU Commission, 2021b), this is provided by IEEE 7000™'s task 10.3 e) 8), which reads as follows: "Through design verification and continued monitoring, determine when the design needs to be modified to accommodate changing contexts, different value priorities, or changes in technical needs, and reiterate the applicable processes" (p. 49).

High-risk-based technical design

Complementing IEEE 7000™'s standard risk-based design path, VBE offers an additional High-Risk-based Design path that builds on and extends the standard path for some systems, threatening human safety, health and wellbeing to such an extent that requirements engineering should be even more rigorous. Take the value of mental health as an example: If new consumer technology like transcranial direct stimulation

(Figure 6.7) were brought to a mass market without being run through a detailed impact assessment-led framework for health effects, there would be a great risk that it might harm people. At least, research suggests so (Wurzman, Hamilton, Pascual-Leone, & Fox, 2016). The technology is therefore a candidate for requiring even more rigorous ethical and risk-based system design than what has been presented so far.

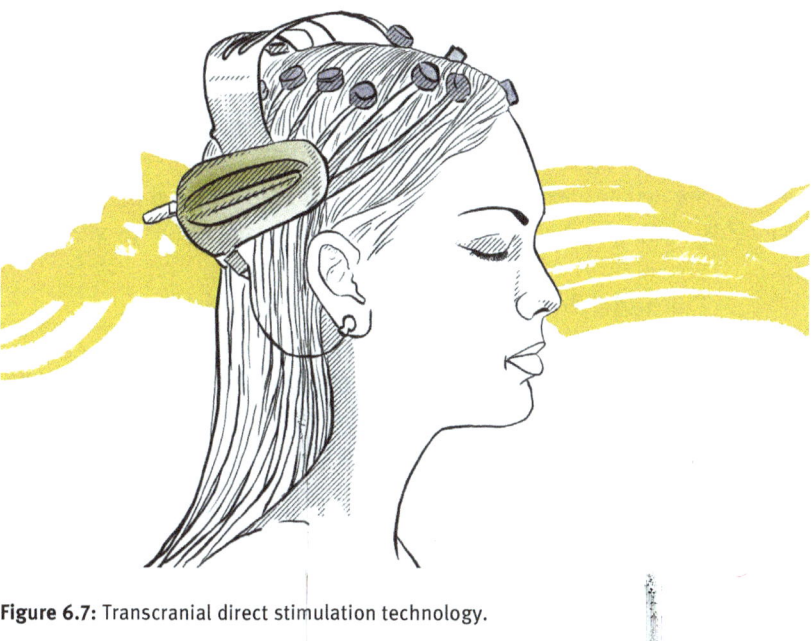

Figure 6.7: Transcranial direct stimulation technology.

A high-risk-based system design path is typically impact-assessment-led. Impact-assessment-led design methodologies are already standardized for safety, security and privacy engineering (for example NIST, 2013; Oetzel & Spiekermann, 2013; Security, 2008). The methods followed in these standards share a common pattern: they all investigate the impact of a technology going astray, and from this analysis, derive a level of protection demand. They engage in a detailed analysis of value threats, and they subsequently align the choice of threat-controls with the level of protection demanded. The bigger the likely negative impact of a technology, the stronger must be the respective technical controls. But before delving into the details of such high-risk-based design steps, a question must be answered: When do organizations actually face such a high risk that an impact assessment-led design process is the best course of action?

When is a system of high-risk nature?

Generally, the extent of negative risk inherent in a system is conceptualized by looking at its potential negative impact (damage potential, adverse effect) as well as the likelihood (probability) of that impact occurring (Black & Baldwin, 2010). The higher the damaging potential consequences of a system and the likelihood of that damage occurring, the higher is its risk. Against this background, the 2006 ISO/IEC 16085 standard on risk management in system and software management defines risk as "the combination of the probability of an event and its consequence" (p. 4 in ISO & IEC, 2006).[66]

However, the two risk dimensions (damage potential consequences and likelihood) also materialize independently. That is, the potential damaging consequences caused by a technology can, by themselves, be so vast that even a small probability (rare likelihood) of their occurrence will not be able to balance the overall risk. An example is nuclear power technology. The damage potential from design faults in nuclear power plants is so severe for people and nature that even the tiniest likelihood of these playing out is not able to offset the overall risk level. Figure 6.8 illustrates this dynamic.[67]

RISK	Insignificant	Minor	Moderate	Major	Severe
Almost certain	Medium	High	High	Extreme	Extreme
Likely	Medium	Medium	High	Extreme	Extreme
Possible	Low	Medium	Medium	High	Extreme
Unlikely	Low	Low	Medium	High	High
Rare	Low	Low	Low	Medium	High

Damage Potential/ Negative Impact/ Adverse Consequences (columns); Likelihood (rows)

Figure 6.8: Risk matrix.

Conceptualizing risk in this way has a high logical appeal. However, the model disguises the true qualitative challenge in judging likelihood and the consequences, appropriately and consistently. In real-world cases, project teams and/or regulators have a hard time correctly determining the damage potential associated with a system failure. Organizations can turn to monetary indicators, of course, to judge damage potential; for example, by looking at the likely damage-removal costs, lost sales, compensation payments or legal penalties impacting the balance sheet, in case the damage occurs. But collecting these numbers is cumbersome. And more importantly, it is much harder to assign such monetary damage tags to the social, human or environmental costs that are also typically associated with a derailing system. Furthermore, the likelihood of damaging events is difficult to anticipate with precision. How can anyone know about the true probability of damage occurrence when there are "unknown-unknowns" and "known-unknowns" that drive the true risk of projects?[68]

Another criticism of this form of risk conceptualization is that it takes a purely organizational perspective. It only asks whether the consequences of a technology are severe for an organization, for instance, in financial terms, while having less regard for human or social harm. Alternatives for risk conceptualization have therefore been proposed, framing risk more explicitly from a human and social perspective, especially in the face of Algorithmic Decision Systems (ADMs), such as AI. A noteworthy approach is taken by Krafft, Zweig and König (2020), who propose that it should be the degree of human and social vulnerability (or severity/intensity of harm) that determines the damage potential of a system. This vulnerability can be understood by looking at the actual values at stake in terms of the number of people affected or by its aggregated collective effects.[69] So, if a system undermined, for instance, the mental health of its users, this would imply a higher vulnerability than if it just undermined their efficiency. Complex information systems like social networks are known for their power to undermine democratic stability, for instance, when they distribute fake news or put users in echo chambers. Such collective effects make whole nation states vulnerable.

The vulnerability dimension is combined in Krafft et al.'s model with a dimension describing the scope of a system's agency. How much scope is there for a system to diverge from the legitimate expectations and interests of affected stakeholders (Krafft et al., 2020)? When a system is highly complex and hard to inspect or largely automated, the scope for humans' agency loss is high. Equally, when the use of a system is imposed by regulators or when users have no alternative choice of a competitive offer, the scope of a system is also high. Krafft et al. argue that the higher the scope of a system is due to its own degrees of freedom and diffusion, the higher is the risk that people and/or society may be harmed.

As the two conceptual approaches to system risk illustrate, there are many variables and debatable criteria to judge as to when exactly a system is one of high-risk. This ambiguity might seduce system providers into arguing that their system is not high-risk, thus saving them the time and money that impact assessment-led system design entails.

In the face of this challenge, regulators like the European Commission have been turning to relatively concrete guidance on when systems are high-risk or not high-risk. The 2021 draft AI regulation of the EU Commission has argued, for instance, that there is always a high risk when the values of health or safety are impacted or when a system impacts values that are endorsed as the EU's fundamental rights (EU Commission, 2021b). The latter are spelled out in the EU Charter of Fundamental Rights and include human dignity, privacy, fairness (non-discrimination), equality between men and women, freedom of expression, and freedom of assembly among others (EU, 2012).

Furthermore, there are concrete technology domains that are per se considered high-risk by the regulator due to their nature, scope, context or purposes of processing. The draft AI regulation of the European Commission lists some of these, including:
- Biometric identification and categorization of natural persons
- Management and operation of critical infrastructure (such as supply of water, gas, heating and electricity)
- Education and vocational training
- Employment, workers' management and access to self-employment
- Access to/enjoyment of essential private services and public services and benefits (like credit and emergency first response services)
- Law enforcement
- Migration, asylum and border control management
- Administration of justice and democratic processes

For all these concrete technology deployment areas, the regulator demands that further concrete values, such as transparency, human oversight (control), accuracy, robustness, security, etc., are respected in the system design (HLEG of the EU Commission, 2020). Taken together, it becomes clear that system providers (at least in Europe) need to take great care of a wide spectrum of values, either because the regulator demands it, or because the context of operation and corporate risk management recommends it.

Organizations that use Value-Based Engineering with IEEE 7000™ for system design are in a good position to comply with this demand. Due to the analyses described earlier, they can show in a transparent and traceable way how they elicited their system's values and know whether their context and concept of operations touches upon any of the values recognized in the law. They also prioritized core system values and, in doing so, have considered regional laws and/or existing industry standards. Moreover, they have performed a threat-control analysis, and therefore have a sound understanding of the uncertainties inherent in delivering relevant value qualities. Thus, they already have a first indication of the level of risk inherent in their system design.[70] The next step for them is to refine and detail the risk-logic they have already pursued for the standard risk-based design path. In other words: VBE organizations needing high-risk-resilient systems first run through the standard risk-based design path outlined above (middle path in Figure 6.3), and then complement it with impact assessment activities.

Impact-assessment-led risk assessment activities are not included in IEEE 7000™ because they have largely been standardized elsewhere. The method presented hereafter, for instance, is similar to the security assessment procedures prescribed by the German Federal Office for Information Security (BSI (Bundesamt für Sicherheit in der

Informationstechnik), 2008) and was derived from the Privacy Impact Assessment standard developed in cooperation with the BSI for the privacy-friendly use of RFID technology (Oetzel & Spiekermann, 2013).

Analyzing the impact of a high-risk system

For impact assessments, project teams start out from the value qualities of the given core value that needs special protection (e.g., safety, privacy, health). Remember that each core value has various value qualities relevant in the system's context. Value qualities were derived from stakeholder concerns in the value exploration phase and they were refined by the Value Lead who consulted the literature or existing legal/industry standards to conceptualize them. Now, *impact* is determined by analyzing in the case of each value quality, how severe the consequences would be for various stakeholders, if it were undermined. Some standards frame the analysis of this "impact" as understanding "protection demand" (BSI (Bundesamt für Sicherheit in der Informationstechnik), 2008) or revealing the "importance" (IEEE, 2021b) of a value quality. The impact (protection demand, importance) can be found by analyzing the effects of what would happen if, for various stakeholders, the value quality were undermined. Figure 6.9 provides a schema for doing this analysis. It distinguishes various stakeholder perspectives, specifies the likely damage scenarios and captures the nature of the anticipated damage. Subsequently, project teams can decide on the extent or severity of the damage that determines protection demand.

The impact analysis schema (shown in Figure 6.9) includes a qualitative evaluation of the damages, in addition to the more quantitative, asset-driven evaluation of some security or safety assessments (see e.g. ISO, 2008), which may be noted in the column called "financial situation." VBE includes context-driven qualitative evaluation dimensions (e.g., mental health or social standing) that the project team must decide on. This is because the consequences of many value quality breaches are of a "soft" context-determined nature. Privacy breaches, for example, often relate to hurt feelings. How would one, for instance, quantitatively assess the impact of a leaked body scan that fully exposed a passenger's figure?

Because it is difficult to quantify the emotional and personal consequences of many human-centric value breaches, VBE distinguishes between limited, considerable and devastating consequences for each damage scenario. Depending on the highest level of consequence identified for any damage scenario, VBE then calls for a corresponding degree of protection demand, which can be low, medium or high. Later, this evaluation (low, medium, high) is used to choose system controls that are commensurate in terms of strength and vigor.

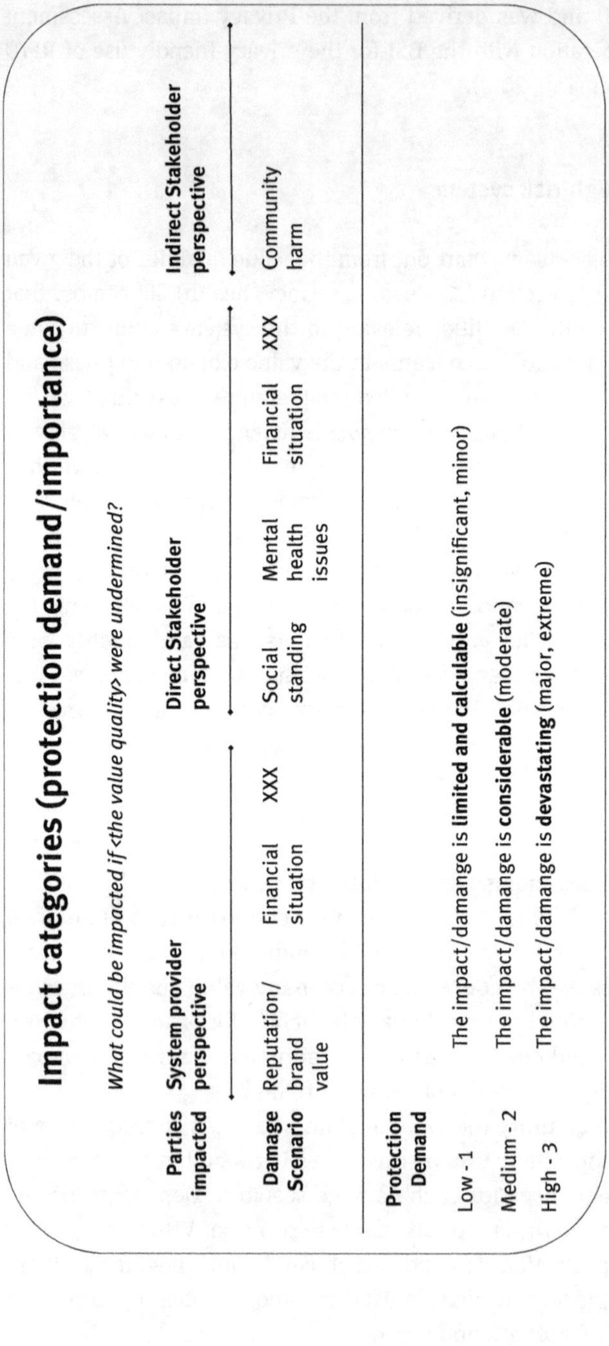

128 — Chapter 6 Value-Based Engineering phase 3: ethically aligned design

Impact categories (protection demand/importance)

What could be impacted if <the value quality> were undermined?

Parties impacted	System provider perspective		Direct Stakeholder perspective			Indirect Stakeholder perspective
Damage Scenario	Reputation/ brand value	Financial situation	Social standing	Mental health issues	Financial situation	Community harm
		XXX			XXX	

Protection Demand

Low - 1 The impact/damange is **limited and calculable** (insignificant, minor)
Medium - 2 The impact/damange is **considerable** (moderate)
High - 3 The impact/damange is **devastating** (major, extreme)

Figure 6.9: Impact assessment of a value quality's breach.

EVR threat analysis for high-risk systems

After analyzing the impact of each value quality, project teams are sensitized for the importance and priority of the assorted values and their related EVRs. The value qualities and EVRs, with a medium or high degree of protection demand, are now scrutinized in more detail than was required for the simple EVR analysis outlined earlier in Figure 6.5. That said, the earlier EVR analysis serves as an input taken by VBE teams to look at the EVR threats in more detail. Not all of these EVR threats are equally likely to occur. The probability of their relevance depends on many factors, such as the IT architecture, the stakeholders involved, the information governance and culture of an organization, the degree of outsourcing of software development, the complexity of the system-of-system partners, the attractiveness and sensitivity of the personal data involved, the education of employees, etc. For example, the probability that a privacy policy is not meaningful and disguises relevant information can depend on the corporate culture.

As many variables influence the likelihood of an EVR threat, security assessments typically require the estimation of concrete threat-*probabilities*. These probabilities, ranging from 0 to 100, determine what threat-controls (system requirements) the development teams are supposed to prioritize later during implementation. If an EVR threat related to security is highly probable, it should also be addressed in the system with an equally high priority in development. The same is the case for other values whose breach would have a direct quantifiable impact on corporate assets.

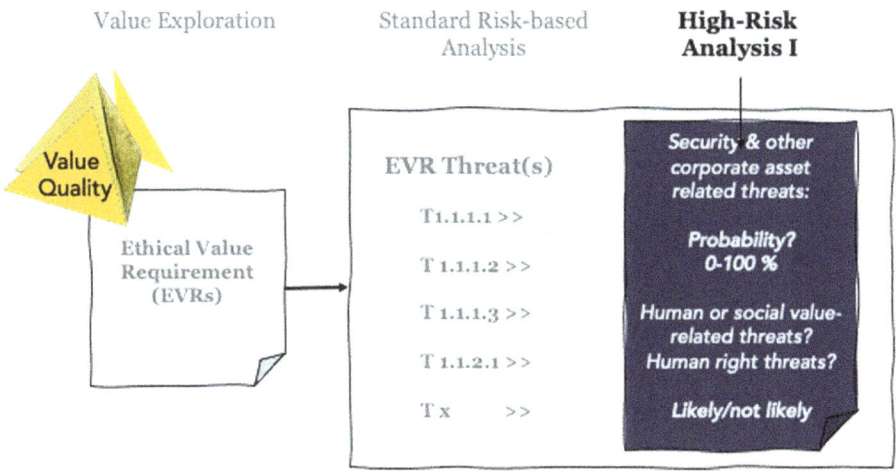

Figure 6.10: Understanding the likelihood of an EVR threat.

However, the estimate of a concrete threat-probability is not advisable for value quality domains that are bound to core human and social values. If human rights, such as fairness or equality, were threatened, for example, they should be treated as early as possible in system development, regardless of their probability score. Therefore, VBE does not assign a threat-probability score when it comes to human rights. A threat is either likely and thereby in play, or it is unlikely and therefore not in play. And if the threat is in play, it must unequivocally be addressed through system design. The control (system requirement) addressing it must be prioritized in system development and become part of the earliest possible product releases. Figure 6.10 summarizes the analysis of EVR threat analysis.

The 2021 legal trial involving Facebook's Instagram service illustrates why VBE is so strict about always prioritizing the control of known human rights-related threats. In this case, the whistleblower Frances Haugen argued that Facebook had been ignoring the known negative impact of Instagram on users' mental health (Figure 6.11). Apparently, the company found through its user research that the Instagram app aggravated negative bodily self-perception for 32% of its female teenage users. But Facebook did not act upon this knowledge. Perhaps, the company thought that a threat probability of 32% would be negligible?[71] Facebook subsequently went on trial for this ignorance of a known human damage potential. It found itself reproached for knowing that its service threatened mental health and for not taking control measures to prohibit this effect. The case shows that when it

Figure 6.11: Frances Haugen accuses Facebook.

comes to human values like mental health, corporate conduct is judged by the public in a black and white manner. Did the company know about the threat? If yes, then they have to do something about it. If not, they are not to blame. VBE would have recommended that Facebook document the researched threat and control for it in the next possible product release.

Choosing the right system controls

The crucial step in an impact assessment-led system design process is to identify controls (system requirements) that can minimize, mitigate or eliminate the identified threats. As outlined for the simple risk-based design process, controls can be technical or non-technical. Technical controls are directly incorporated into the SOI, whereas non-technical controls are management and administrative controls as well as accountability measures.

Controls that are more rigorous and extensive are also likely to be more costly and difficult to realize. For this reason, VBE recommends three levels of control rigor: (1) satisfactory, (2) strong and (3) very strong. For each value quality, the level of system rigor that is required depends on the degree of protection demand that was determined earlier (Figure 6.9). For example, high (3) protection demand, combined with a likely threat, should be mitigated with very strong (3) controls. In contrast, value qualities with low (1) impact can be countered with a satisfactory (1) control. As Figure 6.12 shows, one control can also address and mitigate multiple value threats.[72]

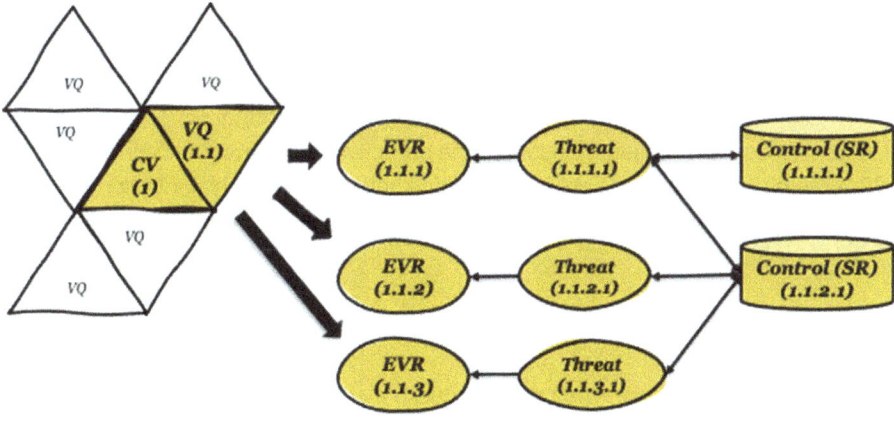

CV = Core Value; VQ = Value Quality; EVR = Ethical Value Requirement; SR = System Requirement

Figure 6.12: One system control can mitigate several threats.

Once the value-based system controls (requirements) that mitigate the value threats are chosen, they must be integrated with the requirements derived from non-value-based design steps. The feasibility of all the recommended controls must be revisited with those system engineers who have driven the non-value-based technology requirements to this point. Also, the organization may again invite stakeholders, in particular, to discuss the acceptability of alternative technical control options. For example, suppose that a retailer wants to use RFID tags on products and has the control alternatives of either killing tags at store exits or deactivating them with a password protection scheme. Let's presume both of these privacy controls would offer the same level of privacy value protection. In this case, the retailer can discuss the control options with engineering teams and external stakeholders, and jointly determine which approach would be most acceptable.

As control alternatives are evaluated and chosen, the organization updates its concept of operations and development plan, clearly identifying how and when each threat is going to be mitigated in the system and also where threats may initially remain unaddressed. After all, not all system requirements can be implemented at once in every case. Threats that remain initially unaddressed (because they are not prioritized for system development) constitute what is called a "residual risk." A residual risk also exists if an implemented control only reduces the impact of a threat but does not eliminate it completely. Whether a residual risk is acceptable depends on the risk management standards of an organization (Naoe, 2008). Residual risks should be noted in the Value Register.

VBE foresees that this whole process of deriving and choosing control requirements, prioritizing them in system development and noting residual risks is documented in the Value Register, described by IEEE 7000™. Here, the trace is kept from core values to respective value qualities, EVRs, threats and controls chosen for implementation. This document ensures that companies can demonstrate to regulators and potential customers, how they acted upon value threats, and can thereby prove their honest efforts to respond to them through the control mitigation plan. The documentation also helps development teams and the organization at large to keep track of its ethical maturity. Everyone is able to see which values are already controlled for, how they have been implemented in the SOI and which ones are still on the waiting list.

Check questions

- How are EVRs different from system requirements?
- What change is induced in an SOI roadmap, once VBE is used?
- How does an impact-assessment-based SOI design differ from a standard risk-based SOI design?
- Why would VBE not assign damage probabilities to high human values/ human rights?

Chapter 7
Transparency and information management

There are several information entities collected for a VBE project in order to demonstrate compliance with IEEE 7000™. These are entered in what is called a "Case for Ethics" (Annex I, p. 75 in IEEE, 2021a), which contains a "Value Register."

The Case for Ethics will be completed in concert with a VBE project.

1. Introduction
   1.1 Societal context
       1.1.1 What market or industry is the SOI in?
       1.1.2 In what countries/cultural regions will the SOI be deployed?
   1.2 Key drivers
       1.2.1 How might the SOI add value (drive) to the societal context in a positive way? / System Purpose

2. SOI, scope and boundaries (Initial concept of operations)
   Note: The high-level concept of operation may need to be sensitive to one or more SOI Views
   2.1 **SOI Description**
       2.1.1 What the SOI is about, what it does and what it aims for
       2.1.2 Block diagram(s) depicting the SOI's internal and external elements
   2.2 **Context**
       2.2.1 Stakeholders:
           2.2.1.1 Direct stakeholders, with various likely roles
           2.2.1.2 Indirect stakeholders, including generic entities such as society and nature
           2.2.1.3 Organizational stakeholders (management representatives, project team)
       2.2.2 Data and service flows:
           2.2.2.1 Contextual (or similar) diagram capturing partners and SOS for which responsibility is required
           2.2.2.2 Data flows (highlighting personally identifiable data flows)
           2.2.2.3 Data controllers and data processors
           2.2.2.4 SLAs and Access Control vis-à-vis partners
       2.2.3 Supporting or dependent systems (SOS)
           2.2.3.1 Boundary choice of SOI analysis
           2.2.3.2 Partner analysis and terms

3. Setting the ethical context outcomes

The real-world context explored and/or the realistic scenario(s) are described, including assumptions potentially made and the potentially diverse SOI Views considered.

3.1 **Realistic scenario description or deployment context**
SOI scenarios of use that are realistic <u>and</u> potentially ethically problematic
   3.1.1 Envisaged market share assumption
   Note: Typically, for an ethical analysis, the assumption is made that the envisioned SOI will be used "at scale" (i.e., will be pervasive and will be a monopoly)
   3.1.2 Assumed place(s) of service usage (industry, households, where stakeholders encounter it)
   3.1.3 Assumed geographic location(s) of service offering
   3.1.4 Assumed primary user interface(s) (if available)

3.2 **Preliminary harms and benefits**
What values are immediately uncovered as essential?

3.3 **Key stakeholders involved in consultation**
Which stakeholders are chosen as representatives?
Can these stakeholders represent outside and critical views?

3.4 **Consultation**
How and when will stakeholders be involved?
Ideal speech situations

3.5 **Value Register**
   3.5.1 Value list
       3.5.1.1 Value Elicitation Matrix accumulated in philosophical mode, Figure 5.10
       3.5.1.2 Value list refined by value quality specifications, Figure 5.12
   3.5.2 Value clusters
       3.5.2.1 Conceptually derived clusters containing the value qualities, aggregated from stakeholders (similar to Figure 5.11 or Figure 5.14)
       3.5.2.2 Conceptually refined clusters documented by the Value Lead (similar to Figure 5.17)
       3.5.2.3 Stakeholder cluster approval signature
   3.5.3 Value narrative
       3.5.3.1 Cluster priority list and choice criteria
       3.5.3.2 Value strategy narrative
       3.5.3.3 Stakeholder value narrative approval signature

4. **Enterprise ethical value-based strategy**
   4.1 Enterprise ethical policy statement (summarizes the value priorities in a public document)
   4.2 EVRs and their value origins, with threshold levels, Figure 6.2 ("EVRs and their origins are traced via a numbering system in the Ethical Value Register")
   4.3 Enterprise-wide ethically aligned design processes (specify which EVRs are only of administrative nature, which ones need a standard risk assessment and which ones need a high-risk impact-based design)

5. **Ethical value risk assessment and management outcomes**
   5.1 Ethical values at risk: evaluation and tolerability criteria
      5.1.1 EVRs and their value origins, with threshold levels, Figure 6.2
      5.1.2 For high-risk systems: "Impact assessment of a value quality's breach," Figure 6.9 as well as a trace of the threat probability in the Value Register become part of Figure 6.5
   5.2 Ethical values sustained or promoted
      5.2.1 EVRs, their threats and the system requirements to control the threats are summarized, Figure 6.5 ("Tracing standard risk-based design in table form")
      5.2.2 System requirements, as identified in Figure 6.5, are brought in line with non-value-based system requirements and prioritized for handling (depending on corporate roadmap formats)
   5.3 Risk mitigation and control options for ethical values at risk
      5.3.1 All EVR threats are matched with the respective system requirement controls
      5.3.2 For high-risk systems, these controls must match the level of protection demand, as identified in Figure 6.9 and traced as part of an extended Figure 6.5

6. **Functional and non-functional requirements traced in the system design**
   Trace how the system features or administrative processes and forms correspond to the requirements

7. **Ethical claims for the SOI and conclusions**
   Summary of core value, value quality, EVR chains per release; documentation of residual risks

8. **Principal resources and references**
   Responsible personnel involved, References used

Chapter 8
Epilogue: dormant values versus gadgetism

Before closing this guide for better engineering, a few reflections can illuminate how VBE is fundamentally different from modern innovation practices and how working with dormant value potentials could trigger a form of positive social change that other innovation standard procedures will have a hard time replicating.

Until the thirteenth century, "innovation" was not a term used. But the phenomenon existed. Indeed, visiting gothic cathedrals today can give a notion of what it means to innovate with Value-Based Engineering. With unmatched technical skill and know-how, craftsmen built edifices that have endured for centuries, and in the elegant, soaring column structures, they expressed that value which they wanted to worship with the buildings: holiness (Duby, 1983). Also, those times were more modest. If there was change, then this was not necessarily marketed with screaming enthusiasm. Change was accepted as incremental and positive, when respectfully embedded in a social context or in a proven existing practice, just as VBE seeks to embed new SOIs respectfully in a context, sensitively reaching out to stakeholders. Finally, progress was not confounded with novelty. Instead, it was deeply human-centric, in that it stood for the maturity of a person's virtues and capabilities.

All this changed in the modern era. Slowly and over the centuries, the idea of human excellence through character formation sank into the world of middle-earth fairy tale, and was replaced by a belief in mastery over nature and of human perfection, whereby perfection nowadays stands more for "having" things (knowledge, cars, etc.) than for "being" a certain kind of a person (Fromm, 1976). The antique conception of human-embedded progress has been replaced by the worshipping of what Jaron Lanier calls "gadgets" (Lanier, 2011) (Figure 7.1).

Today the worshipping of new things has become so pronounced that humanity is involved in a permanent gadget transition. Humanism has gradually been put into these gadgets' service. Tech players herald their inventions to such an extent that progress seems set in stone just because something is digitalized. Only a few individuals dare to doubt that the digital brainchildren might not always bring the positive progress they promise, and that they might instead be often ethically problematic, superfluous and riding on market failure.

Take the idea of General AI. Many people believe that this technology will soon be part of our everyday lives. They imagine confronting a humanoid as depicted in films such as *Terminator*, *Westworld* or *Ex Machina*. They are deeply convinced that such robots WILL come, WILL change the industry (if not the world) and WILL be better than humans. Some countries like Saudi Arabia have given citizenship to these machines, and politicians are led to draft person-rights for them. Recently, a Google developer even hired a lawyer for a chatbot he believed to possess consciousness (Tiku, 2022). Certainly, AI is growing impressively powerful within its natural limits (Spiekermann,

Figure 7.1: Jaron Lanier's earthly thoughts.

2019).[73] However, the "WILL-rhetoric" of technological determinism and the belief in its truth has become so strong that a critical mind must sometimes wonder whether the future can really be predicted – or whether it is just the exceptional personal will of a few visionary figures who deeply believe that it is within the scope of human power to determine the future. Realistically, it is wiser to believe that even the most extraordinary will, coupled with the most abundant money supply, is in reality limited by external constraints – external constraints stemming not only from real technical limitations inherent in the physics of the digital (Spiekermann, 2019, 2020) but also political, economic, social as well as cultural factors, which are recognized in VBE and which strongly determine whether an invention can turn into what we call an "innovation"; that is, an invention seeing productive exploitation and use by people in a market (Hausschildt, 2004).

To understand the real challenges to progress, it makes sense to discern pure-will innovations from real-value innovations. Pure-will innovations may be defined as products and services springing largely from a disembodied human fantasy, inspired by the cultural narrative of a time. Today, this narrative is heavily dominated by transhumanistic science fiction.[74] In contrast, real-value innovations spring from the dormant value potential inherent in the material world of a time. Values like ecological care, conviviality or competence, for instance, drive people across the world today as climate change, loneliness and industry-specialization impact us on an almost daily basis.

Real-value innovations seem to materialize at a point in time when society needs them. Think of today's increasing awareness of the natural environment, and our increasingly concerted efforts to protect it. For a long time, technological inventions or ideas to protect the environment (such as electric cars) lingered in corporate drawers. Though already invented, they did not materialize as innovations. As the context evolved though, this changed. The dormant value of environmental sustainability is rising in importance. And innovations that cater to this reality are called "real", because in contrast to pure-will innovations, they are responsive to the world we are living in. That said, pure-will innovations have their role to play und should therefore be better understood as to their potentials and dangers.

Pure-will innovation

Science fiction stories have been inspiring humanity ever since the second half of the nineteenth century. Jules Verne (1828–1905) and H.G. Wells (1866–1946) wrote about aircraft, high-speed underground travel, space colonization, satellite TV, etc., anticipating a flurry of technologies we are acquainted with today (Figure 7.2). Since the 1950s, Gene Roddenberry, the creator of *Star Trek*, Isaac Asimov, the author of *Robot Tales*, and William Gibson, the inventor of the Metaverse, have had a huge influence on the evolution of digital technology. They are the spiritual founding fathers of mobile communication, robots and cyberspace. Science fiction as an art form leaves deep traces on our technical culture. Their dreams seem to enrich us, mentally as well as physically. However, when dark dreams like the one of General AI, of immortality in a cyborg, of military robots and drone fleets suck away billions in capital investment, the question should arise if that wealth was well spent and if sci-fi dreams continue to be in the service of humanity.

Furthermore, the science fiction narrative provides the intellectual soil for our function-focused and purely technical understanding of progress. It seems to be the embracing of sci-fi narratives that justifies so many technical capabilities that the real world does not actually need. Of course, new technical capabilities are not directly marketed through science fiction. It is rather a very professional marketing discourse that is organized for the pure-will innovations of today's IT industry. This discourse typically sees some fancy acronym (for example, "RFID," "EDGE," "GPT,"

Figure 7.2: Star Trek announced the mobile phone.

"AI/something") or some nebulous term such as "Big Data," or "Cloud" as a starting point. Then, somehow, this term makes it to the Gartner's annual hype-cycle model (Figure 7.3). A mixture of analyst-fueled tech marketing, major industry events, journalist lulling, supply chain pressure, etc. are used to ensure that everyone in the IT industry and even non-specialists have heard about the (supposedly) brand-new thing or service. At the peak of marketing expenditure, it seems as if the technology is so unavoidable (in German "*alternativlos*") that executives feel obliged to invest.[75] Historical examples are WAP services becoming the Mobile Internet, RFID chips replacing barcodes in everyday products, robots replacing hotel receptionists, blockchain technology replacing notaries, etc.

Despite the elaborate marketing campaigns, the true uptake of these pure-will innovations in terms of real sales and return on investment (ROI) is, however, often disappointing. Some, like the WAP technology, fail completely and silently disappear

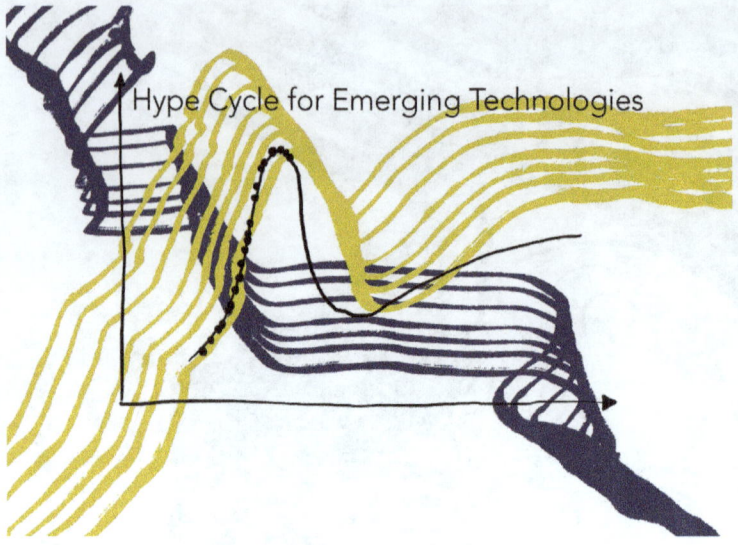

Figure 7.3: Gartner paving the way to market IT with its hype-cycle model.

from the Gartner hypecycle But equally, it is often the case that pure-will innovations drag on. They stick around because they have seen so much financial and human investment put into them that it becomes hard to dispose of them. They may have been integrated so deeply into the corporate IT infrastructure that it is either a risk to dispose of them or it simply feels bad to write them off. After many technical alterations and adaptations, they may have found their place. They completely fail the sci-fiction promises or pathways on the back of which they were initially hyped and invested in. But as the technology is adjusted, as the use-cases are altered, as the market expectations are appeased, etc., the people paid for servicing them get used to them – even live off them. And in the end, the pure-will innovations have influenced the way we do things. It is not that thereby they created human or corporate progress; in truth, they often only created change at a relatively high sunk cost.

This critical account of pure-will innovations is not intended to suggest that science fiction or additional technical capability is bad. High ambitions, like flying to the Moon and Mars, are scientifically inspiring, as are early technical prototypes (Figure 7.4). On the way to realizing tech-dreams, scientists often surpass themselves. As they aim to solve hard problems, they often create unexpected corollary innovations that then turn out to be hugely beneficial. In order to leave Earth, for example, the space industry had to develop highly heat-resistant materials that are now used in car engines. Equally, it needed to develop highly absorptive materials, which are now used in baby diapers. Space travel has also rediscovered the value of hitherto neglected inventions. This is the story of Teflon coating, which is now invaluable for every kitchen. These examples show that the real benefit of pure-will

Figure 7.4: Heart-of-gold rocket.

innovations might be found in secondary inventions that are needed to realize them. In other words, the urge to fulfill a fantasy creates an inner need among innovative minds to solve a problem, which then leads to real-value innovation.

That said, the question remains if better baby diapers and pot coatings can justify the immense expenditure on fantasies when, at the same time, billions of people starve, lack education and access to water and medical services, and while societies are falling far short of the investments needed to protect the very planet we live on.

Finally, what is problematic with sci-fi fantasies is the idea of humanity embedded into them. Science fiction stories share a vision for humanity that is rooted not in individual wisdom, but in technological enhancement (Harari, 2017) and potentially hive-wisdom (see, e.g., the Borg vision in *Star Trek*). Very often, humans are sketched out as Cyborgs – hybrids between a human and a technological species (e.g., *Blade Runner*, *Star Trek*) – or they are portrayed as an energy resource to machines (e.g., *The Matrix*). The visions laid out for the human species rarely describe humans who leverage the potentials of their natural essence (as is the case with Luke Skywalker in *Star Wars*). Instead, they typically portray a world in which a small elite of technology controllers reign over a dumbed-down, degraded and neglected human race. This mainstream portrayal of humans can become a problem because it can incentivize technology providers to make technology choices that shun the power of a human's natural development. Value-Based Engineering seeks to set a counterweight here. It does not recommend envisioning machines as being virtuous, but instead helps innovation teams to tease out the real limitations of digitization. At the same time, it makes projects think about real-value potentials for human beings in the here and now, respecting the essences that stakeholders want to protect and strengthen.

Real-value innovations

Real-value innovations often come into the world in one of two ways. Either they are the response to a pressing need, like the need for vaccines in the face of Covid-19 or the need for a proper rocket coating to fly to Mars. Alternatively, they spring from the individual need of lead users who are drawn to unveil a dormant value they recognize as unfulfilled. An example for a lead user is Gary Fisher, who invented the mountain bike. Gary Fisher was a professional bicycle racer who wanted to go downhill on gravel roads for training purposes. However, bikes in the early 1970s had slim tires and mediocre brakes and so it was impossible to ride down a hilly gravel road. Together with his friend Joe Breezer, Fisher therefore started to modify the slim racing bikes of his time with parts from motor bikes in order to increase their stability and improve the security through better brakes. As a result, the first mountain bike was created and marketed in 1977. In this way, Fisher got access to values relevant to him: higher cycling security, better bike robustness and increased physical joy when riding roads that were formerly inaccessible to bikes (Figure 7.5).

The mountain bike example seems boring when compared to *Star Trek* or *Ex Machina*. But on the other hand an unaccountably bigger number of people have benefited from the intoxicating thrill of going downhill on gravel roads with a mountain bike than traveling to Mars. And the truth is that Gary Fisher is not alone. Lead user innovations, like the mountain bike, are much more common than many experts believe (Bradonjic, Franke, & Lüthje, 2019). In fact, over half (54%) of all relevant innovations in the past decades, across industries, stem from this kind of lead user engagement (Bradonjic et al., 2019). Lead users like Gary Fisher are paving an authentic path to progress, driven by a real bottom-up value aspiration.[76] The question is whether this kind of innovation could actually be systematized.

Figure 7.5: Gary Fisher invents the mountain bike.

Schoolbook innovation management versus real-value innovation

In classical innovation management, as is taught in business schools and Design Thinking schools today, lead users are rarely reflected upon. It is taught that a company typically starts with something tangible. It might start with a prototype or a patent that comes out of R&D. It might seek to leverage external IP it procured, or perhaps it inherits the technology of a company it acquired (Ahmed & Shepherd, 2012). In all these cases, just as in VBE, there is a tangible technology (prototype, IP, etc.); however, unlike VBE, its human or social values are hardly reflected upon. The only motivation for the innovation is to create monetary benefits for shareholders.

Managers typically pursue various strategies to make a technology attractive. In the mid-twentieth century, it would have been enough to simply push the new invention onto the market, but marketers quickly learned that market pull is also important, which requires innovation managers to conduct market research to (ideally) inform product adjustments and marketing messages. In the past 20 + years, market involvement in a product's genesis has considerably matured (Figure 7.6): In fact, when a new technological capability becomes available (time 1), an innovating company now tries to develop it together with customers, from the start. The hugely successful Design Thinking practice (Brown, 2008) organizes this "coupling" of a technology and a market by what is called an "ideation-phase" (time 2). Ideation implies that innovation managers empathize with the direct stakeholders of the envisioned product and try to understand their needs before they sketch out a technology. After ideation, applied R&D and system engineering (time 3) are used to construct the technology in line with the collected market expectations. Prototypes, later called "Minimum Viable Products," which are the result of these efforts (time 4), are built and then further refined in cooperation with users. For instance, a machine learning algorithm might be trained for a hitherto analog purpose. Or a new functional application might be created. Finally (at time 5), the technology is launched and often A/B tested in real time to be continuously improved in an iterative way.

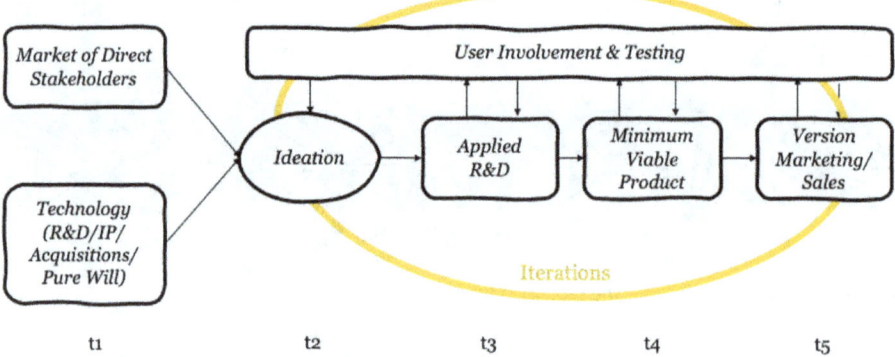

Figure 7.6: Design Thinking has positively influenced how new systems come to market.

Note that the true starting point or source of such an innovation process is very often an already existing technology coming out of R&D, the IP business, or a hype cycle. Alternatively, the source may be a strategic market, customer or technology portfolio analysis, all of which co-determine the motivation of today's corporation to want to invest in a specific technology. In such cases, it is not a genuine need of a person like Gary Fisher or a dormant value that lies at the heart of the innovation project, but rather it is a strategic consideration to secure revenue and/or defend a market position.[77]

In contrast, the initial motivation for real-value innovators is what Plato called the "eros." Eros is far removed from monetary incentives. It is a deep desire for a potential value yet invisible, like the thrill to go safely and wildly downhill on a gravel road with a bike. This is what drives the innovation activity of the lead user (time 1) (Figure 7.7).[78] The eros leads to an intensely creative phase where a first prototype is built, taking whatever technological components are available as a starting point to master the job (time 2). The early prototype is enthusiastically demonstrated to a smaller community of friendly individuals (who are generous enough to overlook the many bugs . . .) (time 3). Grudgingly, but wisely, the lead user sets out to perfect his or her prototype (time 4). The result can be a minimum viable product (if the lead user's motivation persists) (time 5). And following this, word-of-mouth – especially, today, via social media – can make investors jump on the new bandwagon and bring the product to a wider audience, which can then potentially lead to following the traditional innovation process outlined above (time 6). Figure 7.7 summaries this real-value innovation process.

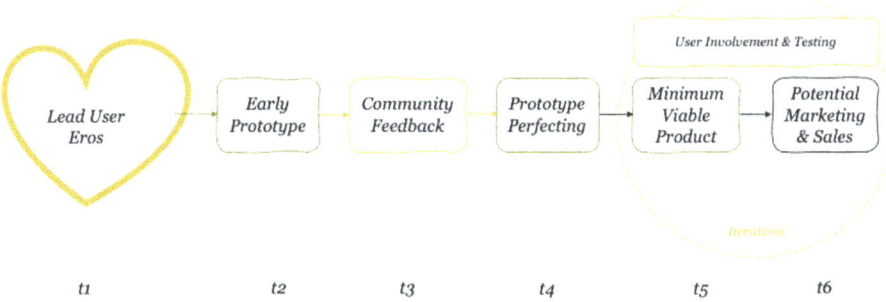

Figure 7.7: The real-value innovation process starts with eros.

Great minds in the field of corporate innovation will of course grasp the importance of real-value innovation, of eros and non-monetary engagement with a thing in its own right. But it is not easy to reproduce such genuine creativity in a corporate environment. An attempt some organizations pursue is to engage in open user innovation processes (Baldwin & von Hippel, 2011). Here, the eros of enthusiastic customers and their creativity is harvested systematically for product evolution. An example of this practice is the "Lego Challenge." Lego regularly invites its customers to participate in contests in which kids and their families can propose their own early prototypes. A community of users and a corporate Lego jury then gives feedback on how to perfect the prototype (Figure 7.8). If a prototype gets significant community support, then Lego builds the proposed new toy and markets it. Presuming that Lego Challenge participants don't primarily engage for the sake of being marketed later, Lego seems to have successfully integrated a real-value innovation flow into its business. Speaking with VBE terminology, customers suggest new value qualities and value dispositions for Lego's building blocks (value bearers).

Note one caveat, however. User innovation in this form can only be used by corporations for product evolution, or what is also called "incremental" or "sustaining" innovation. This is because corporations always need to have a value bearer already at the core of their business mission before they involve users. So, the innovation flow they engage in creates additional value from something tangible that already exists.

Figure 7.8: Lego inspires customers to innovate.

In contrast, disruptive innovations, in the sense described by Clayton Christensen, often create completely new value propositions from something that does not exist yet (Christensen, Raynor, & McDonald, 2015). And herein lies a challenge for incumbent companies. They need to be prepared for disruptions from dormant values yet invisible that alter the markets in which they operate. This is a form of innovation one can hardly expect from lead users.

Dormant values in disruptive innovation

In 1997, Clayton Christensen first coined the term "disruptive innovation" and, in the process, greatly inspired the tech industry. After all, the tech industry sees itself very much in the business of disruption due to its services and products constantly replacing what existed before. But in 2015, Christensen lamented the fact that people tend to call all kinds of new successful businesses "disruptive" even though they are far from what he would consider them to be (Christensen et al., 2015). Uber, for instance, would often falsely (so he argued) be given this title. According to Christensen, technical innovations such as Uber are just sustaining, and not disruptive. They make services or things only slightly different, a little newer or a bit cheaper. But truly disruptive innovations, he argues, should only be heralded as such when a number of criteria are met, which may be seen in the light of what in this chapter is referred to as real-value creation.

An historic example of disruptive innovation, in Christensen's sense, is the printing press. Of course, people used books before the printing press. There was incredibly expensive manual book production. But there was also an initially invisible dormant market segment unserved by this elite book market. The majority of people could not afford handmade books. Disruptive innovations are inserted into an existing practice pursued by incumbents, but they address a hitherto unserved part of the market, such as, in this case, the huge segment of citizens embracing the dormant value potential of suddenly having access to knowledge, religion and education at an affordable price (Figure 7.9).

Figure 7.9: The innovation of the printing press was disruptive because it unveiled dormant value potentials.

In the printing press example, it seems at first sight as if only the lower price led to the success of printed books. But Christensen argues that disruptive innovations do not only originate in low-end product versions of high-end markets; on the contrary, they create entirely "new-market footholds" where none existed before. The printing press, for instance, which came apparently out of nothing, created a completely new foothold of readers. People who were initially illiterate now started reading. People changed due to the availability of technology, addressing a dormant value potential in humanity: the ability to accumulate personal knowledge.

But how can something of such importance be created out of nothing? Christensen gives no answer to this question. In line with the value philosophy promoted in this book, one could argue that a new-market foothold is created because disruptive innovations cause values to flourish that were not actualized before. The value qualities created by Johannes Gutenberg through the printing press, for example, could be described as more regularity, evenness and symmetry in books, thereby increasing their core value of readability. More importantly, the efficiency and speed of printing and the ubiquity of access allowed people to nourish a core value like knowledge, or discover the joy of reading.

The example of photocopiers underscores the same argument. Chester Carlson is known to have been seeking pain relief when he invented the photocopier. He was a patent attorney who suffered from arthritis, which meant that the regular manual patent-copying that his job required was so painful that he needed a practical way to cater to his physical wellbeing. He needed to reduce his pain while saving time and maintaining effectiveness in his work.

Both Gutenberg and Carlson uncovered a dormant value potential for which no tangible product existed before they invented it. Their stories suggest that disruptive innovations not only serve unserved customers at cheaper prices but also cater to the dormant values of an unserved market.

An intriguing finding Christensen reports on is that disruptive innovations are often considered inferior by incumbents. Incumbents do not expect the degree of appreciation their disruptive competitor is met with among the unserved. Christensen argues that the main reason for this phenomenon is that incumbents have optimized their processes to serve their high-end niche and they are not willing to change, especially since they deem their offer to be of higher quality (Christensen et al., 2015). Yet, it may be argued that there is another reason for their underestimation; namely, that they have no rational reason to actually anticipate the forceful value potential inherent in the disrupting offer. With the exception of Gary Fisher, not many saw a need for more safety, reliability and fun when biking. Only Chester Carlson, plagued by the pain of arthritis, seems to have seen the need for an alternative to manual copying. People who are removed from a hypothetical problem do not seem to be in a position to perceive a dormant value potential of a non-existing product for a non-existing market. How should they?

Against this background it is not surprising that IBM's president Thomas J. Watson famously said in the 1940s (Figure 7.10): "I think there is a world market for about five computers." Watson was just rational. But rationality is not the kind of intelligence required to perceive dormant value potentials, such as orderliness, professionalism, commercial flexibility or rapidity, inherent in computers. Watson did not see the dormant values because they effectively did not exist at the time; they were invisible – and rationality can only operate upon the visible. Even when the value of computers

Figure 7.10: Thomas Watson did not recognize the dormant value potential in computers.

first permeated the market, this was not immediately visible until the sales figures clearly captured the phenomenon. At that point, IBM was forced to make a U-turn. A similar path probably happened with book production. Why should fifteenth-century monks have expected a need among the illiterate to have books? That would have been irrational.

Taken together, it could be argued that disruptive innovations have their surprisingly strong and unexpected effect because they do not serve the visible. Instead, disruptive innovations kindle a dormant desire for new value bundles that only a few lead users are able to initially see and uncover.

Real values versus needs

By emphasizing the role of values in innovation, a divergence should be noticed from a popular term usually used in the innovation literature and in practice: user "needs." Typically, user needs are supposed to be at the origin of innovation. When read carefully, many schools of innovation use this term even to explain disruption, including Christensen himself. But scholars should question the meaning of the word "needs." Strictly speaking, the word "need" linguistically implies that a person experiences a lack of something (Schönpflug, 1998). Citizens in the fifteenth century did not, however, lack books, because very few people could even read. Nor did anyone perceive a lack of the Internet in the 1960s. No-one needed a smartphone until the first one came to market. The common underlying aspect in all these disruptive technologies is that that they did not respond to any felt *need*. Instead, they appealed to a dormant value potential – a dormant value potential that did not materialize until the innovation came into the world. Only at that point – that is, when the innovation has already happened – may people developed a need for it.

$$Needs \neq Values$$

$$Values > Needs$$

Against this background, companies should not focus on investigating customer needs when they seek disruptive innovation. Disruption does not come from people lacking something, or what Design Thinkers sometimes call "pain points." "What the eye doesn't see, the heart doesn't grieve over," so the adage goes. Instead, innovators should investigate the dormant stream of potentially unserved human and social values yet unattained that might motivate people to demand a product or service, and ask themselves whether this value potential is strong enough to create a new-market foothold that can replace an existing practice. This is what Value-Based Engineering does in its value elicitation phase.

That said, where needs become important is when trying to understand lead users. Lead users like Gerry Fisher, Johannes Gutenberg or Chester Carlson, who are so important for society's progress as well as for the economy, seem to have suddenly felt an overwhelming need for that which did not yet exist. For them, the act of creation was born of a necessity – a need springing from their own history and milieu, and resonating with an invisible value potential that is ready to be uncovered from the material world, similar to how artists uncover a sculpture from a stone (Vasari, 1987).

What companies might, therefore, be advised to do is nourish lead users. They could support the makers' scene, for example, or support an educational system that discovers and nurtures such individual talent. To do so, however, it is worthwhile investigating who lead users are and what is happening in their act of creation.

Who are lead users?

To understand who lead users are, today's innovation management literature might benefit from Everett M. Rogers' landmark study on the "Diffusion of Innovation," published in 1962 (Rogers, 1995). Studying specifically the diffusion of weed spray, Rogers found that at the very source of innovation, diffusion is a special kind of person he called "the venturesome." The venturesome are not necessarily "scientists" in our current understanding. Neither Gary Fisher, nor Steve Wozniak, nor Johannes Gutenberg were scientists, as the academic world defines them. However, they all shared certain personality traits that Rogers describes very well: The venturesome have a constant interest in new ideas that often leads them and their life path out of the local circle of peer networks. They are able to understand and apply complex technical knowledge, can cope with a high degree of uncertainty and are not necessarily very popular among peers. Steve Wozniak, for example, who was the lead user and innovator at Apple computers, was described as "shy" (Isaacson, 2011) (Figure 7.11).

Figure 7.11: Steve Wozniak was a lead user because he understood the milieu of computer engineering.

What lead users seem to share is a reliable judgment of the value qualities of the things they work with. Johannes Gutenberg, for instance, was a goldsmith who understood the value qualities of the materials he later used for the printing press. Steve Wozniak was an engineer and hacker from youth onwards, which gave him a schooling in arranging components on a motherboard and writing software to integrate them, intuitively knowing which qualities needed to be in place to make a digital circuit work and which not (Figure 7.11).

Lead users operate in what value-phenomenologist Max Scheler has called a "milieu" (see Chapter 3). A "milieu" is the material resonance space effectively experienced in practice by a person when interacting with a material.[79] Working on his printing press, Gutenberg understood the value qualities of tin and lead, materials to which he had become attuned during his apprenticeship. His implicit knowledge of these materials' qualities allowed him to intuit what is possible and what isn't, given the material disposition. "The human milieu is that part of the entire ambient perceptible situation that is possibly effective on a person", wrote value phenomenologist Max Scheler (1921 (2007)) in 1921.

The milieu in which the value qualities of things inform the experimental process should not be equated with the context of a well-organized scientific process that plans and reasons in advance about every possible experimental setup. If we are skilled, ". . . we possess the ability to 'take practical account' of things, which implies an experience of their efficacy . . . It is this same 'practical accounting' which experientially determines our acting . . . and which is itself 'given' . . . but not before, as a 'reason' for them . . ." (p. 140 in Scheler, 1921 (2007)).

Due to this developed skill milieu, a lead user like Wozniak had the practical knowledge of how to best arrange and integrate computer components, which then "experientially" and intuitively determined how he would make the first Apple computer work. What Wozniak or any innovator engages in is a performative act of creation.

Performative acts in corporate innovation

The performative act of creation coming as a gift to the skilled lead users is of a different nature than the extremely detailed, well laid-out, documented, reasoned, explained or proven causal process that many people believe R&D entails. Performative acts of creation are not controllable, measurable, manageable, reasonable or predictable by means of innovation management. And for this very reason, disruptive innovations can also not be enforced or brought to the world by pure will. They cannot be commanded. And they are not the result of mental fantasies.

A scholar who understood this very well, but who still explored how to inspire management practice in its striving for innovation, is the Japanese scholar Ikujiro Nonaka. Nonaka investigated how to explicate and transfer the implicit skill-based knowledge of experts and make it accessible for better innovation (I. Nonaka & Takeuchi, 1995). In a shared context, which Nonaka called "*ba*" (and that Scheler might have agreed to call a shared milieu), a transfer of knowledge takes place between the skilled expert and his or her apprentice. Through a time-intensive process of actual apprenticeship, where corporate innovators learn about the milieu of a craft, they try to learn from experts the implicit value qualities and necessary dispositions inherent in materials and the best practices to deal with them. These value qualities and practices are then explicated and, if possible, translated into tangible product requirements. One of Nonaka's examples for this time-intensive innovation practice is a bread-baking machine. Innovation managers had to physically learn and embody the bread baking practice themselves. In doing so, they learned the specific twists and hand movements made by professional bakers when treating the dough. Only by physically learning the practice were they able to understand how the later perceivable value qualities, such as fluffiness or consistency of the dough, come about; that is, by a specific twist of the hand. And it is this twist that then enabled them to construct the functioning bread-baking machines. They explicated the practice and spelled out what VBE refers to as "value qualities" in VBE, which then informed the functional requirements of a new machine. They engaged in the practice described for phase 1 of Value-Based Engineering above – context and concept exploration.

Summing up

The reflection of the current innovation theory and practice allows for aligning VBE with what is known in innovation management, and enriching what is observed with an appropriate value vocabulary. That is, the lead users live in a specific personal milieu that allows them to resonate with the potential value qualities dormant in the material they redefine. These latent value qualities are then brought into the world by the lead user's skilled manipulation of the materials and processes, creating a technical functionality that serves as a new value disposition. In other words, lead users effectively create new technical dispositions for yet unknown value qualities and core values to materialize. He or she gives the object the form he or she alone already sees. When the resulting innovative object is then later marketed and used, the hitherto latent value also materializes in the eyes of a new customer. In this way, disruptive innovation comes into the world. And these innovations meet a response from a value-embracing market that is very different from the blithe acceptance of today's IT users, who are expected to blindly swallow the pure-will innovations of tech marketing.

Appendix 1
Case study: the rate your teacher app

You own a venture fund and seek to invest in young and innovative companies. An 18-year-old Austrian high-school graduate called Benjamin approaches you for an investment. He thinks that schools and teachers might benefit a lot, if they got digital feedback from their pupils in a similar way to grades that are submitted today on the Internet for all kinds of services. For this purpose, he has programmed a first prototype of an app he calls "*Lernsieg*" (it already runs on iOS and Android) (Figure 8.1). It allows pupils to grade their school and their teachers. Feedback can be provided on a school's general atmosphere, new media and sports facilities, course offerings, library quality and cleanliness. Teachers can be graded on their lecture quality, fairness, preparedness, timeliness, ability to motivate, patience and assertiveness. Pupils can give up to five stars on these dimensions, with 5 stars = very good, 4 stars = good, 3 stars = satisfactory, 2 stars = poor and 1 star = very poor.

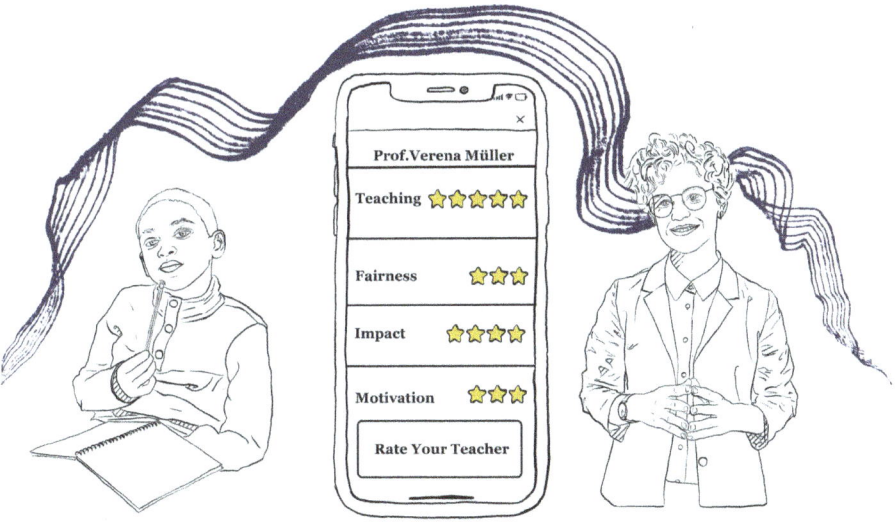

Figure 8.1: Illustration of the Teacher App.

Any pupil from the fifth grade onwards can give her or his feedback. When a teacher receives a grade less than 5 stars, pupils can click on a few preconfigured reasons to explain why they have given the respective grade (see Figure 8.1). But, Benjamin wants to complement these pre-configured choices with a much more sophisticated comment function. In this comment function, pupils should write why they have given a certain grade and be allowed to use as many words as they wish. However, since there are 90,000 teachers in Austria that Benjamin envisions being graded and commented on, the analysis of these comments must be automated. This is one reason why he has approached you as an investor for funding the growth of his service and integrating a sophisticated AI-based text analytics service. Benjamin has found out that there is a US-based AI-service called "Texty" that might become a partner here. Texty presents itself as the market leader in text analysis, able to process any input text and aggregating its content "in a nutshell," including "sentiment analysis" and "synthesizing" text messages and sentiments, across sources. In other words: It would be possible to aggregate all the comments on a teacher collected via feedback (of 2–3 lines in the app's display) containing the presumably most important takeaways on that teacher and the average sentiment shared by the pupils writing the comments.

Benjamin has put forth his ideas in a concept of operations. Here, he also has some additional ideas: comparative teacher and school rankings should become available. He calls these "Stock Charts." Furthermore, teachers and schools should be able to see how they performed over time themselves and in comparison to others. Benjamin also struggles with the problem that he has been forced to manually accumulate all teachers' names from respective school websites, and this only garnered 1,000 teachers' details for his system. He, therefore, wants the Lernsieg teacher database to automatically update itself, when a teacher change is published on a school's public website. What is not recorded so far, and not planned by Benjamin, is the (mostly internal) information on what teacher teaches what student/class. The app only asks for a pupil's age. In other words: the app cannot verify whether a pupil rating a teacher has actually been (or is) in that respective teacher's class (has actually attended his or her lecture).

You inspect the Lernsieg App as it stands. It is easy to use. Pupils do not need to create a proper account when they register themselves with a user name and a password. They just type their mobile phone number into the app and request a verification code. The verification code is sent to the cellphone via SMS, the receipt of which the pupil must confirm. In this way, Lernsieg's server knows that a phone number exists (reacts to the verification code) and that one phone number can only give one rating of a teacher or school and not rate a teacher or school several times. That said, a phone number is able to edit ratings given in the past. The pupil's name (the rater of the teacher) is not displayed to anyone; the pupil is represented through his or her phone number only.

Lernsieg is planning to earn money through advertisement that could be placed in the app, when it is launched. Given that there are over 700,000 pupils in Austria, this might secure reasonable revenue streams.

As an investor, you have committed to only fund start-ups that pass your Value-Based Engineering assessment. You sit down and reflect on whether you should invest or not, against the background of the values impacted by the Lernsieg app as well as the organizational and technical challenges it faces. You derive EVRs and system requirements.

Appendix 2
Notes on the visuals used in this book

If you wanted to draw the abstract concept of "value," what would you draw?

The question of how to depict values as well as their various ontological aspects is a tricky one. For this book, Platon's Timaios dialogue served as an inspiration. Timaios speaks to Socrates, Kritias and Hermokrates about the becoming of the world, the reason for the world's creation and the four elements of fire, water, air and earth that antiquity believed the world to consist of. Timaios also describes what is called the "Platonic solids." These are geometric representations of the four elements and have been influential in architecture; for instance, in the layout of gothic cathedrals, like the cathedral of Chartres.

In this textbook, the use of Platonic solids is only for illustrative purposes and not strictly founded in Plato's philosophy. Other scholars might explore whether the analogies below between Platonic solids, values and technology actually make sense from an academically scrutinized perspective. But a few citations and ideas from Timaios will be shared to playfully reason why, for instance, the choice was made to depict ideal core values as tetrahedra.

The tetrahedron (see Figure 9.1a) in Plato's theory stands for the element of fire. Timaios describes the element of fire, saying: ". . . we must agree that there is first the unchanging idea, unbegotten and imperishable, neither receiving something [aught] into itself from without nor itself entering into something [aught] else, invisible, nor in any wise perceptible" (p. 183 in 52a (Plato, 1888)). Chapter 3 explains how ideal core values like security or beauty are in fact unchanging, imperishable and invisible. Therefore, they have been depicted as tetrahedra in this book, and the element of fire has been attributed to them. Furthermore, fire quite naturally lends itself to the color yellow; so when values or value-laden thought processes are depicted in this book or an inspirational value-rich milieu is at play, the color yellow is used (see Figure 9.1b).

Timaios continues, saying: "Second is that which is named after it [the unchanging idea] and is like to it, sensible, created, ever in motions, coming to be in a certain place and again from thence perishing, apprehensible by opinion with sensation" (p. 183 in 52a (Plato, 1888)). This description expresses, quite essentially, the nature of value qualities, which are moving manifestations of ideal values in perceptible reality, apprehensible by humans and animals. As value qualities are "like" ideas and are "created" and perceivable ("sensible"), they are depicted in this book as the visible faces of the tetrahedron (Figure 9.2).

Timaios then outlines that ". . . every rectilinear plane is composed of triangles . . . These we conceive to be the basis of fire and the other bodies . . ." (p. 191 in 54a (Plato, 1888)). In this book the dotted triangle markers within the tetrahedra faces (see Figure App. 2.2) signal the basis that a perceivable value quality has in the underlying material reality. In Chapter 3, it is outlined, for example, how encryption

Appendix 2 Notes on the visuals used in this book

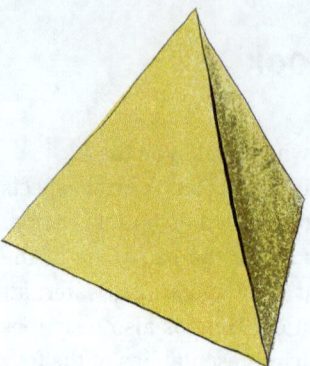

Figure 9.1a: Tetrahedron: value.

Figure 9.1b: Value-inspired thoughts.

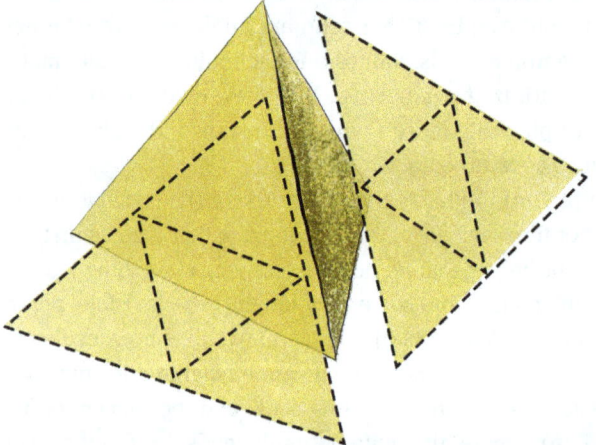

Figure 9.2: Tetrahedron with value quality faces.

would be the basis for the value quality of confidentiality. But encryption, being itself a value disposition, is, in fact, only a potential for confidentiality. It does not guarantee it. If the value quality of confidentiality actualizes due to good encryption, we observe an actualization of a potentiality – and, this is represented here, as a dotted-line triangle.

The Value-Based Engineering (VBE) method shows readers how core values and their value qualities become embedded in software. Software artifacts will, in the end (ideally), manifest or actualize various value qualities that were anticipated in the design process. An octahedron symbolizes software in this book (Figure 9.3). Note, though, that just because an octahedron integrates eight faces (value qualities), software will not necessarily be the accumulation of such a small number of value qualities. The octahedron simply serves as a metaphor, signaling that software should integrate value qualities and the material dispositions that enable them. In Plato's Timaios, the octahedron solid is associated with the element of air.

Timaios says: ". . . two particles of fire combine into one figure of air" (p. 205 in 57a (Plato, 1888)). Are two core values (fire) an ideal number for a software component? This is an open question. Air, or software, is depicted in red in this book.

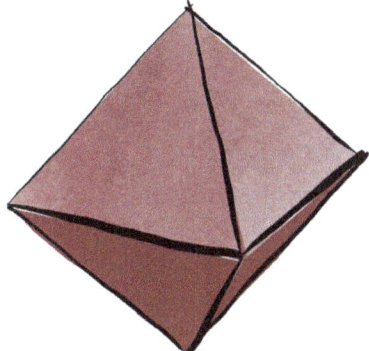

Figure 9.3: Octahedron: software.

Air naturally moves and shapes water. Analogically, software [air] is moving and shaping data. Depending on how software is configured, data is processed differently; so data, as it manifests itself in computer systems, is influenced by the software processing, which it has gone through and will be influenced by the value qualities (solid faces) embedded in the software. For this reason, the icosahedron solid is used to symbolize data in VBE (Figure 9.4a). An icosahedron is the Platonic solid for water. The idea here is also that data flows like water. Furthermore, Timaois says: "when air is vanquished and broken small, from two whole and one half particle [of air] one whole figure of water will be composed" (p. 203 in 57a (Plato, 1888)). Since water is normally depicted in blue, the color blue is used to depict data or facts in this book. Furthermore, whenever processes or decisions are visualized in this book as purely rational or data-driven, blue elements are used in the picture; for instance, when IBM's Thomas Watson calculated the world demand for computers to be three (Chapter 7, Figure 9.4b).

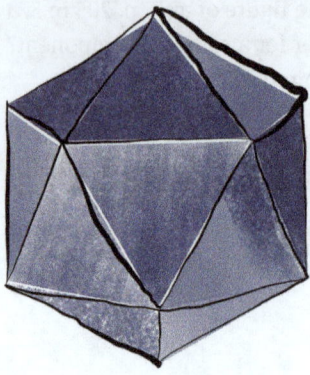

Figure 9.4a: Icosahedron: data.

Figure 9.4b: Data inspired thoughts.

Ideas, software and data come together in one system of interest. A system is symbolized in VBE by a dodecahedron. In Plato's Timaiso, the dodecahedron represents the "universe" (p. 197, in 55 c (Plato, 1888)) (Figure 9.5), and indeed, a system of interest is akin to a small universe. As mixing the colors of yellow, red and blue create green, the systems abstractly represented by the dodecahedron are drawn in green.

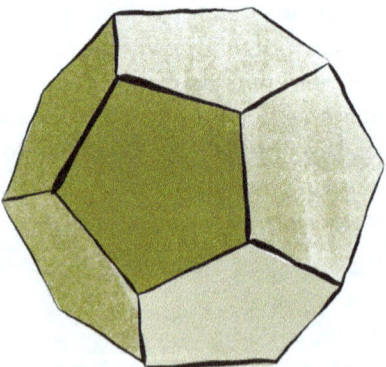

Figure 9.5: Dodecahedron: SOI.

Finally, there is the solid of the cube. Four isosceles triangles (value potentials) form one of six sides of a cube. The cube in Plato's philosophy stands for the element of earth. And, earth is brown. In this book, the earth (cube) is always the starting point for an SOI (Figure 9.6). A visible brown plane (one visible cube surface) is symbolically the "earth" into which new value ideas (fire) are inserted. The element of earth also corresponds to the analogy of a garden alluded to throughout the book. Value ideas are planted into the earth. Mingled with air (software) and water (data) something new is created, or as Plato writes: "When earth meets with fire

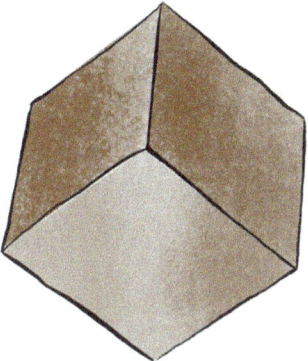

Figure 9.6: Cube: raw earth at the start of a project.

and dissolves by the keenness of it, it would drift about, whether it were dissolved in fire itself, or in some mass of air or water, until the parts of it meeting and again being united became earth once more" (p. 203 in 56d (Plato, 1888)).

References

Aboujaoude, E. (2012). *Virtually You – The Dangerous Powers of the E-Personality*. New York, London: W.W. Norton & Company.

Ahmed, P., & Shepherd, C. (2012). *Innovation Management*. New York: Financial Times/Prentice Hall.

Arendt, H. (1965 (2006)). *Eichmann in Jerusalem: A Report on the Banality of Evil*. New York: Penguin Classics.

Aristotle. (2000). *Nichomachean Ethics* (R. Crisp, Trans.). Cambridge: Cambridge University Press.

Aumayr-Pintar, C., Cerf, C., & Parent-Thirion, A. (2018). *Burnout in the Workplace: A Review of Data and Policy Responses in the EU*. Retrieved from Luxembourg.

Baldwin, C. Y., & von Hippel, E. A. (2011). Modeling a paradigm shift: From producer innovation to user and open collaborative innovation. *Organization Science, 22*(6), 1399–1417. doi:10.1287/orsc.1100.0618.

Bednar, K., & Spiekermann, S. (2021). On the power of ethics: How value-based thinking fosters creative and sustainable IT innovation *Working Paper:* https://bach.wu-wien.ac.at/d/research/results/ris/export/97385/.

Bednar, K., & Spiekermann, S. (2022). Eliciting values for technology design with moral philosophy: An empirical exploration of effects and shortcomings. *Science, Technology, & Human Values, forthcoming*.

Bednar, K., Spiekermann, S., & Langheinrich, M. (2019). Engineering privacy by design: Are engineers ready to live up to the challenge? *Information Society, 35*(3), 122–142. doi:10.1080/01972243.2019.1583296.

Berenbach, B., & Broy, M. (2009). Professional and ethical dilemmas in software enginneering. *IEEE Computer,* (1), 74–80. doi:10.1109/MC.2009.22.

Black, J., & Baldwin, R. (2010). Really responsive risk-based regulation. *Law & policy, 32*(2), 181–213. doi:10.1111/j.1467-9930.2010.00318.x.

Boehm, B. W. (1988). A spiral model of software development and enhancement. *IEEE Computer Society Press, 21*(5), 61–72. doi:10.1109/2.59.

Bradonjic, P., Franke, N., & Lüthje, C. (2019). Decision-makers' underestimation of user innovation. *Research Policy, 48*(6), 1354–1361. doi:10.1016/j.respol.2019.01.020.

Brown, T. (2008). Design thinking. *Harvard Business Review, 86*(6), 84–92.

BSI (Bundesamt für Sicherheit in der Informationstechnik). (2008). Risk analysis on the basis of IT-Grundschutz. In Bonn.

Cambridge Dictionary. (Ed.). (2014). Cambridge Dictionary. Cambridge University Press.

Catelluccia, C., & Le Métayer, D. (2019). *Understanding algorithmic decision-making: Opportunites and challenges*. Research Report.

Christensen, C. M., Raynor, M. E., & McDonald, R. (2015). What is disruptive innovation? *Harvard Business Review, 93*(12), 44–53. Retrieved from https://hbr.org/2015/12/what-is-disruptive-innovation

Christl, W. (2017). *Corporate Surveillance in Everyday Life – How Companies Collect, Combine, Analyze, Trade, and Use Personal Data on Billions*. Retrieved from Vienna: http://crackedlabs.org/dl/CrackedLabs_Christl_CorporateSurveillance.pdf

Cooper, R. G. (2008). Perspective: The stage-gate ideal-to-launch process – Update, what's new, and nexgen systems. *Journal of Product Innovation, 25*(3), 213–232. doi:10.1111/j.1540-5885.2008.00296.x.

Duby, G. (1983). *The Age of the Cathedrals: Art and Society, 980–1420*. Chicago: University of Chicago Press.

Eco, U. (2004). *Die Geschichte der Schönheit*. München, Germany: Carl Hanser Verlag.

https://doi.org/10.1515/9783110793383-011

Charter of Fundamental Rights of the European Union, (2012).
Artificial Intelligence Act, COM(2021) 206 final C.F.R. (2021a).
Proposal for a regulation of the European parliament and of the council laying down harmonised rules on artificial intelligence (artificial intelligence act) and amending certain union legislative acts, COM(2021) 206 final C.F.R. (2021b).
Regulation (EU) 2016/679 of the European parliament and of the council of 27 April 2016 on the protection of natural persons with regard to the processing of personal data and on the free movement of such data, and repealing Directive 95/46/EC (General Data Protection Regulation), L 119/1 C.F.R. (2016).
Regulation (EU) 2016/679 of the European parliament and of the council of 27 April 2016 on the protection of natural persons with regard to the processing of personal data and on the free movement of such data, and repealing Directive 95/46/EC (General Data Protection Regulation), L119/1 C.F.R. (2016).
Feser, E. (2013). Kurzweil's phantasms – a review of how to create a mind: The secret of human thought revealed. *First Things*. Retrieved from https://www.firstthings.com/article/2013/04/kurzweils-phantasms
Frankena, W. (1973). *Ethics* (2nd ed.). New Jersey, USA: Prentice-Hall.
Frauenberger, C. (2019). Entanblement HCI The Next Wave? *ACM Transactions on Computer Human Interaction (TOCHI)*, *27*(1), 1–27. doi:10.1145/3364998.
Friedman, B., & Hendry, D. (May, 2012). *The Envisioning Cards: A Toolkit for Catalyzing Humanistic and Technical Imaginations*. Paper presented at the Computer Human Interaction (CHI), Austin, Texas, USA.
Friedman, B., & Hendry, D. (2012b). *The envisioning cards: a toolkit for catalyzing humanistic and technical imaginations*. Paper presented at the Conference on Human Factors in Computing Systems, New York, NY, USA.
Friedman, B., & Hendry, D. G. (2019). *Value Sensitive Design: Shaping Technology with Moral Imagination*. Mit Press.
Friedman, B., & Kahn, P. (2003). Human values, ethics, and design. In J. Jacko & A. Sears (Eds.), *The Human-Computer Interaction Handbook* (pp. 1177–1201). Mahwah: NY, USA: Lawrence Erlbaum Associates.
Fromm, E. (1976). To have or to be. New York: Harper & Row.
Fuchs, T. (2017). *Ecology of the Brain*. Oxford: Oxford University Press.
Fuchs, T. (2020). *Verteidigung des Menschen – Grundfragen einer verkörperten Anthropologie* (2. Auflage ed.). Berlin: Suhrkamp Verlag.
Fusaro, R. (2004). None of our business. *Harvard Business Review*, *82*(12), 33–44.
Gibson, J. J. (1986 (1979)). *The Ecological Approach to Visual Perception*. Hillsdale, NJ: Erlbaum.
Gotterbarn, D., & Rogerson, S. (2005). Responsible risk analysis for software development: creating the software development impact assessment. *Communications of the Association of Information Systems*, *15*, 730–750. doi:10.17705/1CAIS.01540.
Habermas, J. (1985). *The Theory of Communicative Action Vol 1: Reason and the Rationalization of Society*. London: Beacon Press.
Hammad, M., Inayat, I., & Zahid, M. (2019). *Risk Management in Agile Software Development: A Survey*. Paper presented at the 2019 International Conference on Frontiers of Information Technology (FIT), Islamabad, Pakistan.
Hao, K. (January 21st, 2019). AI is sending people to jail – and getting it wrong. *MIT Technology Review*. Retrieved from https://www.technologyreview.com/2019/01/21/137783/algorithms-criminal-justice-ai/
Harari, Y. N. (2017). *Homo Deus: A Brief History of Tomorrow*. Harper Perennial.
Hartmann, N. (1932). *Ethics*. London: George Allen & Unwin.

Hartmann, N. (1953). *New Ways of Ontology* (R. C. Kuhn, Trans.). Chicago: Henry Regnery Company.

Hausschildt, J. (2004). *Innovationsmanagement* (3rd ed.). Munich, Germany: Verlag Vahlen.

HLEG of the EU Commission. (2020). *Assessment List for Trustworthy AI (ALTAI)*. Retrieved from Brussels: https://ec.europa.eu/newsroom/dae/document.cfm?doc_id=68342

Hobbs, D. (2017). *Investigations of Worth: Towards a Phenomenology of Values*. (PhD). Marquette University, Milwaukee, Wisconsin. Retrieved from https://epublications.marquette.edu/cgi/viewcontent.cgi?article=1751&context=dissertations_mu

Hoff, J. (2021). *Verteidigung des Heiligen – Anthropologie der digitalen Transformation*. Freiburg: Herder.

Hoffer, J. A., George, J. F., & Valacich, J. S. (2002). *Modern Systems Analysis and Design* (3rd ed.). New Jersey, USA: Prentice Hall.

Husserl, E. (1908–1914 (1988)). Vorlesungen über Ethik und Wertlehre. In U. Melle (Ed.), *Husserliana Band XXVIII*. Dodrecht: Kluwer Academic Publishers.

IEEE. (2021a). IEEE 7000 – Model Process for Addressing Ethical Concerns During System Design. In. Piscataway: IEEE Computer Society.

IEEE. (2021b). IEEE Standard for an Age Appropriate Digital Services Framework Based on the 5Rights Principles for Children. In. New York: IEEE Consumer Technology Society.

Ihde, D., & Malafouris, L. (2019). Homo faber revisited: Postphenomenology and material engagement theory. *Philosophy & Technology, 32*, 195–214. Retrieved from https://link.springer.com/article/10.1007/s13347-018-0321-7

Introna, L. (2009). Ethics and the speaking of things. *Theory, Culture & Society, 26*(4), 25–46. doi:10.1177/0263276409104967.

Isaacson, W. (2011). *Steve Jobs: A Biography*. New York: Simon & Schuster.

ISO. (2008). ISO/IEC 27005 Information technology – Security techniques – Information Security Risk Management. In (Vol. ISO/IEC 27005:2008): International Organization for Standardization.

ISO. (2014). ISO/IEC 27000 Information technology – Security techniques – Information security management systems – Overview and vocabulary. In (Vol. ISO/IEC 27000:2014): International Organization for Standardization.

ISO. (2015). ISO/IEC/IEEE 15288: Standard on Systems and software engineering – System life cycle processes. In (Vol. ISO/IEC/IEEE 15288:2015(E), pp. 1–118). Geneva, New York: ISO/IEC.

ISO, & IEC. (2006). Systems and software engineering – Life cycle processes – Risk management. In (Vol. ISO/IEC 16085:2006(E)). Geneva: Software & Systems Engineering Standards Committee of the IEEE Computer Society.

ISO/IEC. (2018). ISO/IEC 29101: Information Technology – Security techniques – Privacy architecture framework. In I. I. J. S. 27 (Ed.), (Vol. ISO/IEC 29101:2018(E)): DIN Deutsches Institut für Normung e.V.

ISO/IEC/IEEE. (2017). ISO/IEC/IEEE 12207: 2017, Systems and software engineering – software life cycle processes. In.

Jobin, A., Ienca, M., & Vayena, E. (2019). The global landscape for AI ethics guidelines. *Nature – Machine Intelligence, 1*, 389–399. doi:10.1038/s42256-019-0088-2.

Kant, I. (1785/1999). Groundwork for the metaphysics of morals (M. J. Gregor, Trans.). In M. J. Gregor & A. W. Wood (Eds.), *Practicle Philosophy*. New York: Cambridge University Press.

Kelly, E. (2011). *Material Ethics of Value: Max Scheler and Nikolai Hartmann* (Vol. 203). Heidelberg London New York: Springer.

Klenk, M. (2019). Moral philosophy and the 'ethical turn' in anthropology. *Zeitschrift für Ethik und Moralphilosophie, 2*, 331–353.

Klenk, M. (2021). How do technological artefacts embody moral values. *Philosophy & Technology, 34*, 525–544. doi:10.1007/s13347-020-00401-y.

Kluckhohn, C. (1962). Values and value-orientations in the theory of action: An exploration in definition and classification. In T. Parsons, E. A. Shils, & N. J. Smelser (Eds.), *Toward a General Theory of Action* (pp. 388–433). Cambridge, Massachussetts: Transaction Publishers.

Krafft, T. D., Zweig, K. A., & König, P. D. (2020). How to regulate algorithmic decision-making: A framework of regulatory requirements for different applications. *Regulation & Governance*. doi:10.1111/rego.12369.

Krasnova, H., Widjaja, T., Buxmann, P., Wenninger, H., & Benbasat, I. (2015). Research note – why following friends can hurt you: An exploratory investigation of the effects of envy on social networking sites among college-age users. *Information Systems Research*, 26(3), 585–605. doi:10.1287/isre.2015.0588.

Laakasuo, M., Drosinou, M., Koverola, M., Kunnari, A., Halonen, J., Lehtonen, N., & Palomäki, J. (2018). What makes people approve or condemn mind upload technology? Untangling the effects of sexual disgust, purity and science fiction familiarity. *Nature*, 4(84). doi:10.1057/s41599-018-0124-6.

Lahlou, S., Langheinrich, M., & Röcker, C. (2005). Privacy and trust issues with invisible computers. *Communications of the ACM*, 48(3), 59–60. doi:10.1145/1047671.1047705.

Lanier, J. (2011). *You are not a Gadget: A Manifesto*. London: Penguin Books.

Le Dantec, Christopher A., Erika Shehan poole, and Susan P. Wyche, 2009. "Values as Lived Experience: Evolving Value Sensitive Design in Support of Value Discovery." Computer Human Interaction (CHI) Conference, Boston, MA, April 7th 2009.

Liao, S. (2018). Amazon warehouse workers skip bathroom breaks to keep their jobs, says report. *The Verge*. Retrieved from https://www.theverge.com/2018/4/16/17243026/amazon-warehouse-jobs-worker-conditions-bathroom-breaks

MacIntyre, A. (1984). *After Virtue: A Study in Moral Theory* (2nd ed.). Notre Dame, Indiana: University of Notre Dame Press.

Makena, K. (2019). Google hired microworkers to train its controversial Project Maven AI. Retrieved from https://www.theverge.com/2019/2/4/18211155/google-microworkers-maven-ai-train-pentagon-pay-salary

Maslow, A. (1970). *Motivation and Personality* (2nd ed.). New York: Harper & Row Publishers.

Mingers, J., & Walsham, G. (2010). Toward ethical information systems: The contribution of discourse ethics. *MIS Quarterly*, 34(4), 833–854. doi:10.2307/25750707.

Mumford, E. (2000). A socio-technical approach to systems design. *Requirements Engineering*, 5(2), 125–133. doi:10.1007/PL00010345.

Munn, L. (2020). Angry by design: Toxic communication and technical architectures. *Humanities and Social Sciences Communications*, 7, 53. doi:10.1057/s41599-020-00550-7.

Nagel, T. (1992). *Der Blick von Nirgendwo*. Frankfurt/Main Suhrkamp.

Naoe, K. (2008). Design culture and acceptable risk. In P. E. Vermaas, P. Kroes, A. Light, & S. A. Moore (Eds.), *Philosophy and Design – From Engineering to Architecture* (pp. 119–130). Miton Keynes UK: Springer Science + Business Media.

NIST. (2013). *NIST 800-53: Security and Privacy Controls for Federal Information Systems and Organizations*. Gaithersburg, MD: U.S. Department of Commerce.

Noe, A. (2005). *Action in Perception*. Cambridge: MIT Press.

Nonaka, I., & Takeuchi, H. (1995). *The Knowledge Creating Company: How Japanese Companies Create the Dynamics of Innovation*. London: Oxford University Press.

Nonaka, I., & Takeuchi, H. (2011). The wise leader. *Harvard Business Review*, 89(5), 58–67.

Norman, D. A. (1988). *The Psychology of Everyday Things*. New York, USA: Basic Books.

Oetzel, M., & Spiekermann, S. (2013). A systematic methodology for privacy impact assessments: a design science approach. *European Journal of Information Systems*, 23(2), 126–150. doi:10.1057/ejis.2013.18.

Orlowski, J. (Writer). (2021). The Social Dilemma. In E. Labs (Producer).
Osterwalder, A., & Pigneur, Y. (2010). *Business Model Generation: A Handbook for Visionaries, Game Changers, and Challengers*. Hoboken, New Jersey, USA: John Wiley & Sons.
Plato. (1888). *The Timaios of Plato*. New York: MacMillan and Co.
Polanyi, M. (1974). *Personal Knowledge: Towards a Post-Critical Philosophy*. Chicago: University Of Chicago Press.
Porter, M., & Kramer, M. R. (2011). Creating shared value. *Harvard Business Review, 89*(1–2), 62–77.
Pruitt, J., & Grudin, J. (2003). *Personas: Practice and Theory*. Paper presented at the Conference on Designing for user experiences (DUX'03), San Francisco, California, USA.
Rogers, E. (1995). *Diffusion of Innovations* (4th ed.). New York, USA: The Free Press.
Ronnow-Rassmussen, T. (2015). Intrinsic and extrinsic value. In I. Hirose & J. Olson (Eds.), *The Oxford Handbook of Value Theory* (pp. 29–43). New York: Oxford University Press.
Ross, W. D. (1930). *The Right and the Good*. Oxford: Oxford University Press.
Sagoff, M. (1986). Values and preferences. *Ethics, 96*(2), 301–316. doi:10.1086/292748.
Scheler, M. (1921 (1973)). *Formalism in Ethics and Non-formal Ethics of Values: A New Attempt Toward the Foundation of an Ethical Personalism*. USA: Northwestern University Press.
Scheler, M. (1921 (2007)). *Der Formalismus in der Ethik und die Materiale Wertethik – Neuer Versuch der Grundlegung eines ethischen Personalismus* (2. unveränderte Auflage ed.). Halle an der der Saale: Verlag Max Niemeyer.
Scholtes, J. (August 17th, 2015). Price for TSA's failed body scanners: $160 million. *POLITICO*.
Schönpflug, U. (1998). Bedürfnis. In *Historisches Wörterbuch der Philosophie* (Vol. 1, pp. 765–771). Deutschland: Schwabe Verlag.
Schwab, K. (2017). *The Fourth Industrial Revolution*. New York: Currency.
SDSN. (2022). *World Happiness Report 2022*. Retrieved from New York: http://worldhappiness.report/
Technical Guidelines for the Secure Use of RFID, (2008).
Shanafelt, T., West, C., Sinsky, C., Trockel, M., Tutty, M., Satele, D., . . . Dyrbye, L. (2019). *Changes in Burnout and Satisfaction With Work-Life Integration in Physicians and the General US Working Population Between 2011 and 2017*.
Shilton, K. (2013). Values levers: Building ethics into design. *Science, Technology & Human Values, 38*(3), 374–397. doi:10.2307/23474474.
Skelton, A. (2012). William David Ross. In *The Stanford Encyclopedia of Philosophy*. Stanford: The Metaphysics Research Lab.
Sommerville, I. (2016). *Software Engineering* (10th ed.). International: Addison Wesley Pub Co Inc.
Spiekermann, S. (2012). Privacy-by-design and airport screening systems. Retrieved from https://www.derstandard.at/story/1331779737264/privacy-by-design-and-airport-screening-systems
Spiekermann, S. (2016). *Ethical IT Innovation – A Value-based System Design Approach*. New York, London and Boca Raton: CRC Press, Taylor & Francis.
Spiekermann, S. (2019). *Digitale Ethik – Ein Wertesystem für das 21. Jahrhundert*. Munich: Droemer.
Spiekermann, S. (2020). Human intelligence vs. artificial intelligence: On the unethical effects of false anthropomorphism. In A. C. f. R. a. T. Development (Ed.), *Ethical Challenges in the Age of Digtial Change*. Wien.
Spiekermann, S. (2021a). Value-based Engineering: Prinzipien und Motivation für bessere IT Systeme. *Informatik Spektrum, 44*, 247–256. Retrieved from https://link.springer.com/article/10.1007/s00287-021-01378-4?wt_mc=Internal.Event.1.SEM.ArticleAuthorAssignedToIssue&utm_source=ArticleAuthorAssignedToIssue&utm_medium=email&utm_content=AA_en_06082018&ArticleAuthorAssignedToIssue_20210831

Spiekermann, S. (2021b). Zum Unterschied zwischen künstlicher und menschlicher Intelligenz und den ethischen Implikationen der Verwechselung. In K. Mainzer (Ed.), *Philosophisches Handbuch Künstliche Intelligenz* (pp. 1–20). München: Springer Verlag.

Spiekermann, S., Korunovska, J., & Langheinrich, M. (2018). Inside the organization: Why privacy and security engineering is a challenge for engineers. *Proceedings of IEEE, 107*(3), 1–16. doi:10.1109/JPROC.2018.2866769.

Spiekermann, S., Winkler, T., & Bednar, K. (2019). A telemedicine case study for the early phases of value based engineering. In I. f. I. a. Society (Ed.), (Vol. 1). Vienna: Vienna University of Economics and Business.

Taatgen, N. A., & Lee, F. J. (2003). Production compilation: A simple mechanism to model complex skill acquisition. *Human Factors, 45*(1), 61–76. doi:10.1518/hfes.45.1.61.27224.

Tiku, N. (June 11th, 2022). The Google engineer who thinks the company's AI has come to life. *The Washington Post*. Retrieved from https://www.washingtonpost.com/technology/2022/06/11/google-ai-lamda-blake-lemoine/

Transatlantic Reflection Group. (2021). In Defence of Democracy and the Rule of Law in the Age of "Artificial Intelligence". In T. F. Society (Ed.). Online: Transatlantic Reflection Group, on Democracy and the Rule of Law in the Age of "Artificial Intelligence".

Ulrich, W. (2000). Reflective practice in the civil society: The contribution of critically systemic thinking. *Reflective Practice, 1*(2), 247–268. doi:10.1080/713693151.

Vallor, S. (2016). *Technology and the Virtues – A Philosophical Guide to a Future Worth Wanting*. New York: Oxford University Press.

van de Poel, I. (2018). Design for value change. *Ethics and Information Technology, 23*, 27–31.

van de Poel, I., & Kroes, P. (2014). Can technology embody values. In P. Kroes & P.-P. Verbeek (Eds.), *The Moral Status of Technical Artefacts* (pp. 103–124). Delft: Springer.

Vasari, G. (1987). *Lives of the Artists* (Vol. 1). New York: Oxford University Press.

Venkatesh, V., Morris, M. G., Davis, G. B., & Davis, F. (2003). User acceptance of information technology: Toward a unified view. *MIS Quarterly, 27*(3), 425–478. doi:10.2307/30036540.

Verbeek, P.-P. (2016). Toward a theory of technological mediation – a program for postphenomenological research. In J. K. Berg, O. Friis, & R. C. Crease (Eds.), *Technoscience and Postphenomenology: The Manhatten Papers* (pp. 184–204). London: Lexington Books.

von Ehrenfels, C. (1890). Über Gestaltqualitäten. *Vierteljahresschrift für wissenschaftliche Philosophie, 4*, 249–292.

Wittgenstein, L. (1993). Cause and effect: Intuitive awareness. In J. C. Klagge & A. Nordmann (Eds.), *Philosophical Occasions* (pp. 1912–1951). Indianapolis: Hacket.

Wurzman, R., Hamilton, R. H., Pascual-Leone, A., & Fox, M. D. (2016). An open letter concerning do-it-yourself users of transcranial direct current stimulation. *Annals of Neurology, 80*(1), 1–4. doi:10.1002/ana.24689.

Zuboff, S. (2018). Big other: Surveillance capitalism and the prospects of an information civilization. *Journal of Information Technology, 30*, 75–89. doi:10.1057/jit.2015.5.

Zuboff, S. (2019). *The Age of Surveillance Capitalism: The Fight for a Human Future at the New Frontier of Power*. New York: Public Affairs.

Endnotes

1 A potentially wrongful assumption of the narrow economic view of value is that wellbeing, sustainability of natural resources, etc. are always convertible into monetary value. This is however difficult, because high values such as freedom or dignity are priceless and can hardly be traded.

2 According to the website https://techjury.net/blog/virtual-reality-statistics/#gref, in 2020 54.8 million US citizens used VR, of which 28% daily.

3 https://www.theguardian.com/lifeandstyle/2019/aug/21/cellphone-screen-time-average-habits.

4 In 2020, for instance, 36 billion records were exposed in the period until Q3 in the US alone: https://pages.riskbasedsecurity.com/hubfs/Reports/2020/2020%20Q3%20Data%20Breach%20QuickView%20Report.pdf.

5 https://www.upguard.com/blog/cost-of-data-breach.

6 https://itchronicles.com/information-security/cyber-security-statistics-2020/.

7 With the use of mechanical and electromechanical machines, we had gotten used to the fact that they can break, but as long as they are not broken, they work very reliably and in a largely error-free way. We mistakenly assume that this is the same with digital systems. It is not surprising that a large part of the development of software systems is spent on an activity called "debugging." For example, in classical aviation, software engineers spend only three out of 24 months writing new program code. The rest is used for testing and debugging. The result of such debugging processes is an official error rate, which is a quality feature of a software. When developing a system for high-security areas, such as in the field of aeronautical engineering or hospital systems, manufacturers aim to get less than 0.5 errors per 1,000 lines of code in each case. And now there are also methods, such as the model-based design of software, which make the achievement of such low error rates easier and faster. But this reassuringly low error rate should not fog our judgment of reality; that is, that complex systems contain millions of lines of code. A highly digitized car, for instance, can contain up to 100 million lines of code. 0.5 errors per 1,000 lines of code, therefore, means 50,000 errors.

8 Between 2014 and 2015 the number of yearly worldwide filed patents increased from 2,7 to 2,9 million according to the World Intellectual property organization: http://www.wipo.int/portal/en/ (27.07.2022).

9 Various sources for this number, ranging from 25% to 83% (being at least partially or totally unsuccessful), are quoted in https://www.ganttic.com/blog/why-do-projects-fail-miserably (27.07.2022).

10 https://www.investopedia.com/articles/personal-finance/040915/how-many-startups-fail-and-why.asp (27.07.2022).

11 Using the word "Design" is ambiguous. There is a design-phase in a technical engineer's system development life cycle. This is not what graphic designers, architects, artists, etc. would understand by the term, though. They likely understand "design" as an accumulation of concretized ideas and sketches, while an engineer would understand by "design" a much more detailed machine model (such as a UML activity diagram or a process model).

12 Important in such a difficult and tense decision space is to maintain an inner sense of what Evagrius Ponticus (345–399) would have called "Apatheia"; that is, to not feel tempted to be drawn to one side from the very start, but to listen and try to understand what a valuable path can be in this context, honestly weighing the views.

13 The Friedman doctrine is still discussed among many businesses still pursuing its guidance in practice. However, even conservative media have started to turn away from it. The *Economist* said in 2016 that a focus on short-term shareholder value has become "a license for bad conduct, including skimping on investment, exorbitant pay, high leverage, silly takeovers, accounting shenanigans and a craze for share buy-backs, which are running at $600 billion a year in America."[7] In 2019,

influential business groups such as the World Economic Forum and the Business Roundtable updated their mission statement, leaving behind the Friedman doctrine in favor of "stakeholder capitalism"[20] (at least on paper if not in widespread practice[21]).

14 Note that the term "blocks" is used here. One could also talk about three "stages" or "phases" of system engineering. However, these terms are avoided, because system engineering was for a long time dominated by a sequential "system development life cycle" thinking, such as the waterfall model, which is now perceived as too rigid and cumbersome for technology projects. Instead, highly iterative and agile forms of system analysis, design and implementation have become the industry norm, which go away from the kind of sequence thinking that is signaled by words like "stages" or "phases."

15 Note that value qualities are called "value demonstrators" in IEEE 7000™.

16 Note that the term "Conceptual Analysis" has been recognized as important by Value Sensitive Design scholars like Batya Friedman (Friedman & Kahn, 2003), who has promoted and used this analysis for a long time.

17 This last process of validation, monitoring and iteration is not well elaborated on in IEEE 7000™. Here, Value-Based Engineering clearly diverges from the standard by providing more guidance on risk-based design for highly sensitive systems.

18 In his review article on "Entanglement HCI," Chris Frauenberger cites (Introna, 2009) writing: "Humans and things are 'ontologically inseparable from the start'" (Frauenberger, 2019).

19 "Artefact x embodies value V if x affords to a set of subjects S in conditions C an ability A and there is reason to positively respond to A (positive value), or there is reason to negatively respond to A (negative value)" (p. 535 in Klenk, 2021).

20 The IEEE 7000™ standard defines a value disposition in line with this view as "a system characteristic that is an enabler or inhibitor for one or more values" (p. 23 in IEEE, 2021a).

21 Mark Sagoff neatly clarifies how using the word "subject" or "intersubjectivity" is not equal to an individual (subjectivist) conception. He writes: "When you and I perceive the same table, for example, our perceiving, being acts of states of mind, are subjective; that which is perceived, the 'content' of the act of perception, however, exists objectively, as what we both see. Thus, even though acts of perceiving are subjective, the object perceived is intersubjective, belonging to a world that is not mine or yours but ours in an epistemological sense" (p. 314 in Sagoff, 1986).

22 Moreover, note that value quality Gestalten are mental objects of consciousness that actualize potentialities of the physical entities that bear them. They are ontologically on a higher "layer" than the value dispositions. Those who have philosophically studied the layered ontology of life have shown how reality is constituted of different layers and how higher layers bear properties that lower layers do not possess (Hartmann, 1953). Take the example of the corpus of a fly that Wittgenstein expands upon. He makes clear that any normal living being is able to recognize that the corpus of a dead fly is different from a live fly. The difference being that the value of "life" actualizes potentialities of the underlying matter, which cannot be weighed, touched, combined or made to cause anything in a Newtonian way. Life actualizes an additional "layer" to the underlying matter as Nicolai Hartmann puts it, and that layer, while depending on the properties of the material layer underneath, still contains properties that the underlying layer does not have (Hartmann, 1953).

23 Note that these mental value quality wholes or phenomena available to human consciousness are not physical. As Edmund Husserl wrote: "Value is not a being, value is something relating to being or not-being, but it belongs in another dimension" (p. 340 in Husserl, 1908–1914 (1988)). Regardless of their real constitution, however, both of these value layers are objectively given. Even though the value layers are not physically touchable or visible, we have given names to values. And how could humans give names to the non-existent? However, the way they are given is as "pre-theoretic reason" in a judgment situation (p. 59 in Hobbs, 2017). Only if need be they can be "re-understood" (German verb: "nachverstehen") (p. 65 in Hobbs, 2017). Note that there are also scientific positions in cognitive neuroscience and philosophy that claim that the world does not really exist, but is created purely in our brains as a kind of simulation. From this perspective, values would be imagined by individuals. Thomas Metzinger, for example, wrote in 2009 that "our brains generate a world-simulation," or Francis Crick (1994) claimed that "[w]hat you see is not really there, it is what your brain believes is there." This picture of a bodiless and worldless subject stuck in an anthropomorphic ego-tunnel is still quite influential; especially since it is actively nourished by science fiction stories like *The Matrix* or *Transcendence*. And as shown in Chapter 7, science fiction actively fuels IT investments and is therefore quite influential. VBE does not follow this neuro-constructivist perspective. There are three reasons for this: First, constructivism seems to be scientifically contested through neuroscience itself as well as through ecological psychology and cognitive science (for a good overview, see Fuchs, 2017; Noe, 2005). Secondly, even though the term "speciesism" is contested, it seems arrogant to put humans into the position of creators of Earth by being-through-their-brains. VBE is based on Max Scheler's *Material Value Ethics,* where he says, ". . . the ego is neither the point of departure for the apprehension nor the producer of essences" (Scheler, 1921 (1973)). Thirdly, it seems that the constructivist perspective on reality is also an extremely dangerous one to follow in times of enormous environmental sustainability challenges. The scientific position that the world is just a simulation would give a justification for humanity to not care for it, since it is imagined anyway. Neuroconstructivism releases humanity from its responsibility towards nature.

24 The discernment of values and value qualities is taken from Scheler, who wrote (p. 13 in Scheler, 1921 (1973)): "Goods and Values – No more than the names of colors refer to mere properties of corporeal things – notwithstanding the fact that appearances of colors in the natural standpoint come to our attention only insofar as they function as a means for distinguishing various corporeal, thing-like unities – do the names of values refer to mere properties of the thing like given unities that we call goods. Just as I can bring to givenness a red color as a mere extensive quale, e.g., as a pure color of the spectrum, without regarding it as covering a corporeal surface or as something spatial, so also are such values as agreeable, charming, lovely, friendly, distinguished, and noble in principle accessible to me without my having to represent them as properties belonging to things or men. Let us first attempt to demonstrate this by considering the simplest of values taken from the sphere of sensory agreeableness, where the relation of the value-quality to its concrete bearer is no doubt the most intimate that can be conceived. Every savory fruit always has its particular kind of pleasant taste. It is therefore not the case that one and the same savor of a fruit, e.g., a cherry, an apricot, or a peach, is only an amalgamation of various sensations given in tasting, seeing, or touching. Each of these fruits has a savor that is qualitatively distinct from that of the others; and what determines the qualitative difference of the savor consists neither in the complexes of sensations of taste, touch, and sight, which are in such cases allied with the savor, nor in the diverse properties of these fruits, which are manifested in the perception of them. The **value-qualities**, which in these cases 'sensory agreeableness' possesses, are authentic qualities of a value itself. And insofar as we have the ability to grasp these **qualities**, there is no doubt that we can distinguish fruits without reference to the optical, tactile, or any other image except that given by taste; of course it is difficult to effect such a distinction without the function of scent, for example, when we are accustomed to such a function. For the amateur it may be difficult to distinguish red wine from white while in the dark. However, this and many similar facts, such as decreased ability to distinguish among flavors when scent is set aside, show only the very many degrees of competence found among the men in question and their particular habituation to the ways in which they take and grasp a particular flavor." In this phenomenological reflection, Scheler first likens the nature of values to the nature of colors. This is helpful, because the comparison allows us to grasp the metaphysical difference between a value bearer with "mere properties" and the value itself: Just because green grass turns yellowish in a hot summer does not mean that the colors green and yellow lose their independent existence as the soil dries. The waves meeting your retina are altered in structure as the soil changes, but the independent understanding of color and the ability to discern colors remains untouched. Moreover, just like for colors, one can close one's eyes and bring values in front of the inner eye. A value is "in principle accessible to me without my having to represent them as properties . . . ." This possibility of accessing values in thought is important for the moral anticipation or impact assessment of values resulting from technology. Scheler then goes on and takes the exemplary value of "sensory agreeableness" of a fruit, which "possesses" "authentic qualities," making the ontological distinction pursued in VBE (introduced through the value example of security). He explains that if one has three different cherry trees and one agrees that all of them taste good (are "sensor[il]y agreeable"), it is still possible to recognize that all three types of cherry taste slightly differently. For one cherry tree the sensory agreeableness of its cherries will come particularly from the quality of sweetness, while for the other cherry tree the quality of juiciness might stick out. "Each of these fruits has a savor that is qualitatively distinct from that of the others . . ." However, since it is a human being who tastes the three different kinds of cherry in all cases and this human being's "complexes of sensations of taste, touch, and sight" are not altered, the difference in the juiciness and sweetness between the cherries (the qualities) can obviously not stem from that human judge or "subject-pole." The differences in taste also do not come from the human eater inspecting all "the diverse properties of these fruits" and then judging the cherry as juicy, because he or she recognizes a relatively high percentage of water in the fruit. Instead, the human eater puts the cherry in his or her mouth and immediately has a taste of the various qualities at a higher layer of consciousness than if he or she discerned the property details (the value dispositions). So, the human experience integrates the various qualities that allow one to draw conclusions on the overall sensory agreeableness of the cherries. For cherries this integration is easy. For technology, and a value like security, expertise is needed to get "a taste" of the various qualities.

25 In computer science, ontologies don't deal with the real nature of being but try to represent conceptions of being in a reduced, man-made and machine-readable metadata structure. This metadata structure (called Resource Description Framework) contains logical relationships between human pre-defined data entities that a computer is able to process, which is only the case if these relationships are provided in the modeling and programming language the computer understands. An example is the Friend-of-a-Friend (FOAF) ontology, which allows machines to integrate family and friend relationships in its reasoning.

26 James Gibson, the inventor of the term "affordance," famously wrote: "The affordances of the environment are what it offers the animal, what it provides or furnishes, either for good or ill" (p. x in (Gibson, 1986 (1979). Future research could investigate what parts of the value ontology are actually best capturing the notion of affordances, which have been associated with values (Klenk, 2019). Are affordances of an ideal value nature embodying an overall value Gestalt? Are they that which the human animal responds to – that is, the value qualities? Or are affordances value dispositions (properties, features) that engineers can actually work on, and which determine how the object can be used, which is what Don Norman, who made the concept of affordance popular in technology design, would argue (Norman, 1988).

27 Hartmann and Scheler were criticized for their value conception by Heidegger, who insisted that any value analysis of being should be more anchored in "the being of the thing" itself. From Heidegger's perspective, "an analysis of the valuative character of valuable things [can only be] a subsidiary enterprise to a thorough inquiry into the being of those objects themselves; values would be nothing more than abstractions incorrectly derived from our actual experience of objects" . . . For Heidegger, any attempt to think of values on their own terms necessarily substitutes an unjustified abstraction for a genuine analysis of the being of valuable things. As he writes in the Humanism letter: "Rather, it is important finally to realize that precisely through the characterization of something as 'a value' what is so valued is robbed of its worth . . . [Valuing] does not let beings: be. Rather, valuing lets beings: be valid – solely as the objects of its doing" (cited from p. 32 in Hobbs, 2017). One could argue that by understanding the process of valuation as one holistic phenomenon by which value dispositions in the thing bear value qualities actualizing ideal values we can overcome the philosophical dispute. Reality is multilayered (for a more nuanced discussion of this philosophical position, see Hoff, 2021).

28 One should note the difference between lead users and Design Thinkers, because lead users are typically domain experts and visionaries that bring a deep milieu experience and technical knowledge to an innovation effort that allows them to grasp dormant value potentials. This is a different starting point than Design Thinking where such lead users are probably not normally present.

29 It could be argued here that the interface to Design Thinking is best when a project has not yet started to engage in building the minimum viable product. An advanced prototyping effort to build a minimum viable product already puts project teams deeply into a creation process leading to a product that they fall in love with even though the technical context has not yet been fully explored as VBE foresees it. For example, the SOS and partner space, the legal and ethical feasibility, the indirect stakeholders, etc. have not been investigated yet. Therefore it is advisable that the prototype should not be set up beyond an initial mock-up.

30 To use the analogy of a garden: When you have a garden with certain flowers that grow in a given soil (water level, pH value, etc.) and you want to re-engineer this garden, wanting, perhaps, to plant different flowers and trees, then you need to understand first whether this is possible; whether the soil, water and weather conditions allow you to do this. You might find out that your new garden vision does not work with the physical conditions you have. Alternatively, you might find out that your existing garden did not work the way you wanted it to work because the conditions were never considered.

31 It is a difference whether one plants a fourfold garden in the city of an Arabic town or whether one plants vegetables in France.

32 A challenge is that many current management tools talking about values equate these with needs or ask for needs when looking for values. This appears to me to be a logical fallacy.

33 https://www.statista.com/statistics/204123/transmission-type-market-share-in-automobile-production-worldwide/ (last visited September 10th 2022) .

34 https://www.forbes.com/sites/neilpatel/2015/01/16/90-of-startups-will-fail-heres-what-you-need-to-know-about-the-10/?sh=2819568c6679 (last visited September 10, 2022) .

35 One might wonder whether the existence of a technology in VBE does not contrast with the last chapter of this book on Real Value Innovations. There it is I outlined that disruptive innovations driven by lead users start with eros and not with a technology. Instead, the personal desire and need of a lead user and his or her ability to uncover hidden value potentials drives a project. This urge of the innovator (not the market!) means that something yet unseen comes into the world. While this seems to be a true phenomenon of how innovation often works, it does not seem to be a process that can be managed or enforced by a company. The eros is a gift. It cannot be brought about by force. VBE as a process and method must build on what is reliably there and what can serve as a reliable and repeatable process trigger. This is typically an SOI of some sort or a first concept of operation.

36 Note that, in contrast, Design Thinking has often been portrayed as a method which first screens the world for needs. This difference might be small, but is important: Value-Based Engineering does not screen the world with the ambition of identifying problems, pain-points or needs solvable with technology. Instead, it assumes that the world is already saturated with values. Given a new technology, however, VBE wonders whether that which is already so abundant can still be positively enhanced with value. Of course, it can happen that a VBE project also identifies negative values unfolding in the real world and sees a chance for using the SOI to mitigate these existing negatives. This is very important as well of course. The world is full of misguided projects that created the very pain-points we now need to get rid of again. But it is important to understand the difference in attitude between VBE and Design thinking: whether you believe that reality is insufficient and in need of a technology or whether you believe that the current playing field is saturated and well-balanced, and therefore necessarily not needing technology.

37 In a brownfield project, an "operational concept" will already exist that sketches out in much more detail how a system is already set up and how data flows within it. This detailed documentation can be used to reconstruct the big picture view that a concept of operation offers for an initial feasibility, stakeholder- and context analysis.

38 For a description of UML component diagrams, see e.g.: http://www.agilemodeling.com/style/componentDiagram.htm.

39 For a fine instruction video on how to draw context diagrams see Karl Wiegand's online tutorial: https://www.youtube.com/watch?v=iY7xZ8Nut5A&list=PLA1dXT4tBFfcRj7WmtSbIMlhKHWWUuktk&index=9 (last visited: September 10th 2022).

40 To determine whether an external service qualifies as an AI service, it is worthwhile considering the definition of AI as it is embedded in the EU Commission's 2021 draft regulation on AI, which says that the use of AI is dependent on the use of certain techniques and approaches "for a given set of human-defined objectives, generate outputs such as content, predictions, recommendations, or decisions influencing the environments they interact with" (p. 39 in EU Commission, 2021b). These techniques are "(a) Machine learning approaches, including supervised, unsupervised and reinforcement learning, using a wide variety of methods including deep learning; (b) Logic- and knowledge-based approaches, including knowledge representation, inductive (logic) programming, knowledge bases, inference and deductive engines, (symbolic) reasoning and expert systems; (c) Statistical approaches, Bayesian estimations, search and optimization methods" (Annex 1 in EU Commission, 2021b).

41 "Personas are archetypal users of or stakeholders in a system. They represent the needs of a larger group in terms of their goals, expectations, and personal characteristics. Personas act as stand-ins for real stakeholders and thus help to guide decisions about system functionality and design targets" (Pruitt & Grudin, 2003 referenced in p. 221 in Spiekermann, 2016).

42 *Forbes*: "50 Stats Showing Why Companies Need To Prioritize Consumer Privacy": 84% of consumers say they want more control over how their data is being used. (Cisco); 81% of consumers say the potential risks they face from data collection by companies outweigh the benefits. (Pew Research Center); 79% of Americans are concerned about how their data is being used by companies. (Pew Research Center); 78% of consumers are most protective of their financial data. (RSA); and 92% of Americans are concerned about their privacy when they use the Internet (TrustArc) (taken from: https://www.forbes.com/sites/blakemorgan/2020/06/22/50-stats-showing-why-companies-need-to-prioritize-consumer-privacy/?sh=1094c03737f6; last visited September 10.

43 This is how a user of IEEE 7000[TM] can understand activity 7.3. g) "Identify and resolve gaps and discrepancies between the assumptions and outcomes of the value-based ConOps and alternative ConOps descriptions" (p. 38).

44 It is contested to what extent machines can have what are called morals. There are of course papers and proposals with titles such as "Moral machines," but conceptually it is dubious as to whether objects without consciousness can have what we call "morals" in a human sense.

45 What is right or wrong may be derived from our understanding of good and bad. It is therefore subsidiary to goodness.

46 Note that the definition of IEEE 7000[TM] is slightly different. In the standard, ethics is defined as "a branch of knowledge or theory that investigates the correct reasons for thinking that this or that is right." With this definition IEEE 7000[TM] makes its reader think that it is strongly rooted in morality, but this is not true of the Material Value Ethics approach. VBE overcomes the shortage of the standard's definition, by adapting the definition in line with the standard's nature.

47 Research has furthermore shown that stakeholders are aware of these dynamics. When asked to imagine the ethical effects of an SOI operated at monopolistic scale, they identify significantly more value issues than if they don't work on this assumption (Winker xxx). **Isn't it Winkler??**.

48 Note that the exact text in the standard for virtue ethical elicitation is slightly different (Section 8.3 b) p. 40):" Conduct a detailed and critical analysis of how the SOI or features within the SOI potentially change user character (virtue-ethical analysis), identifying the potential damage to the character of individual stakeholders that can occur if the system were implemented at scale . . .".

49 Note that the exact text in the standard for utilitarian value elicitation is slightly different (Section 8.3 b) p. 40): Conduct a detailed benefits and harms-based value analysis (utilitarian ethics) as follows: i) Identify benefits for individual stakeholders that can be provided by the SOI if the system were implemented at scale. ii) Identify harms for individual stakeholders that can be caused by the SOI if the system were implemented at scale . . .".

50 Note that the exact text in the standard for utilitarian value elicitation is slightly different (Section 8.3 b) p. 40): Conduct a detailed and critical analysis of how the SOI or features within the SOI potentially challenge the perceived ethical duties of the stakeholders . . .

51 *Apatheia* (Greek: ἀπάθεια; from a "without" and pathos "suffering" or "passion"), in Stoicism, refers to a state of mind in which one is not disturbed by the passions. It is best translated by the word equanimity, rather than indifference. The meaning of the word apatheia is quite different from that of the modern English apathy, which has a distinctly negative connotation. According to the Stoics, apatheia was the quality that characterized the sage (taken from Wikipedia, November 2, 2021, URL: https://en.wikipedia.org/wiki/Apatheia). The term apatheia was originally coined by the desert father Evagrius Ponticus. "For Evagrius apatheia is the goal of monastic ascesis. Apatheia is a state of integration, where enemies cannot trouble, where anxiety cannot disturb, where injury is met with patience, where the changes and chances of mortality do not shake, where the will is detached and unwavering because it is set on God" (Richard Byrne "Cassian and the Goals of Monastic Life." In: *Cistercian Studies Quarterly* 22 (1987), 3–16, 11; taken from Johannes Hofff, "Verteidigung des Heiligen," Herder, 2021, p. 395/396).

52 The cluster-figures can be printed, spread out, have their implications discussed, and can be shifted around by workshop participants and corporate leaders until a final order is agreed upon.

53 The example implies that profitability may indeed be a core value identified in a VBE value elicitation process. If a company is open to publishing that profitability is its most important value over and above any other value, such as equality, trust or dignity, then it may very well do that. Corporate leaders have to sign and publish this prioritization ranking, though, and make it accessible to everyone. The question is whether any executive would want that.

54 The IEEE 7010 standard contains a great list of sources for industry-related value commitments.

55 IEEE 700™ defines a value register as "[an information store created for transparency and traceability reasons, which contains data and decisions gained in ethical values elicitation and prioritization and traceability into ethical value requirements" (p. 23).

56 In IEEE 7000™ this integration is done in Section 9.3 b). IEEE 7000 specifies (p. 45): "Validate the EVRs along with other stakeholder requirements in cooperation with selected stakeholders, including top management and the project team."

57 Note that ISO/IEC/IEEE 15288 defines systems requirements as including functional, performance, process, non-functional, and interface requirements, including design constraints. Only through the word "non-functional" might there be a hint to management measures. VBE and IEEE 7000 diverges from this perspective.

58 An example of a socio-technical measure is the specification of service-level agreements with partners.

59 In 9.3 d) 2) and 3) IEEE 7000 outlines: "2) Analyze and harmonize the EVR and value-based system requirements with requirements derived from non-value-driven means, identifying and rationalizing competing or supportive requirements for the SOI. 3) Analyze EVR value-based system requirements in conjunction with the requirements derived from non-value-driven means for technical and organizational control over the system" (pp. 45–46).
60 A specific solution a technical team might already have decided on may include external partners and components (such as AI skills sourced from outside the SOI) or architecture decisions. If such decisions are already taken, then VBE comes too late and conflict can arise.
61 The EU's draft AI regulation foresees, for instance, in its Article 9 that such a risk management function is set up (see p. 46 in EU Commission, 2021b).
62 Some experts argue that risk "management" includes/comprises the kind of risk-logic or threat-control analysis I describe here in addition to the project management effort. But I prefer to distinguish between management and risk-based system design because typically the kind of people involved in the two tasks are working in different organizational departments. Risk managers often are more on the administrative side of things, while risk-based design requires the involvement of system and software engineers. This does not, of course, exclude that within organizations system and software engineers can also often be risk managers. But the essence of the work to be done is different: one looks after the management of a project, the other looks after what is being put into the technical roadmap concretely.
63 https://www.wu.ac.at/ec/projects/privacy-brochure-a-benchmark-study-1.
64 https://people.cs.kuleuven.be/~koen.yskout/icse15/catalog.pdf.
65 This recording of value-derived system requirements is also likely to be demanded by the AI draft regulation, which asks organizations building high-risk AI systems to show the "intended purpose" of a system. The "purpose" of a system is, however, not its functionality, but the values pursued through a system. For example, the whistleblower Frances Haugen accused Facebook of prioritizing profit over safety. How can an organization prove that it did not prioritize profit over safety? The regulator is likely to expect organizations to illustrate how they embedded and/or prioritized certain aims in a system over others.
66 I am aware that there is a newer (2021) ISO/IEC/IEEE version of this standard that defines risk the way IEEE 7000 does, as the "effect of uncertainty on objectives."
67 The depiction is adapted from https://www.supportadventure.co.nz/risk-management/risk-management-processes/.
68 https://en.wikipedia.org/wiki/There_are_known_knowns.
69 "The severity of harm crucially depends on the nature of decision-making – what is decided upon and what are possible decision-outcomes. ADM systems used for consumer recommendations affect individuals' welfare markedly less than ones used for job recruitment or medical interventions. It furthermore matters how many individuals are affected by the decision-making. Even marginal adverse effects due to an ADM system can amount to significant harm caused if a large number of individuals are affected. Also, some ADM systems may produce aggregate, collective adverse effects that cannot easily be reduced to individual impact" (p. 11 in Krafft, Zweig, & König, 2020).
70 Note that risk in IEEE 7000TM is defined in terms of "uncertainty," or more precisely as the "effect of uncertainty on objectives" (3.1 "risk" p. 20) mentioned in note lxi.
71 https://www.theguardian.com/technology/2021/sep/14/facebook-aware-instagram-harmful-effect-teenage-girls-leak-reveals.
72 It is important also to see the whole control landscape (as in Figure 6.11) complementing the rigor of single controls. Layered levels of weaker controls can sometimes be more effective than a single strong control: "Thousand blades of grass can break the hardest stone." The flow of setting system controls has been embraced by and was already described for privacy impact assessment

driven designs in Oetzel, M., & Spiekermann, S. (2013). A systematic methodology for privacy impact assessments: a design science approach. *European Journal of Information Systems*, 23(2), 126–150.

73 The natural limits of the digital fabric has been discussed in Spiekermann, (2019) and includes the error proneness of software systems in general, and the problems always associated with data quality. When it comes to AI, the Big Data illusion, the problem of meaning, the lack of embodiment, the lack of a socially embedded autonomy, the lack of intelligence in terms of nous, etc. (this has all been discussed in Feser, 2013; Fuchs, 2020; Spiekermann, 2021b).

74 Pure-will innovation can be defined as products and services springing from a disembodied human fantasy that is in line with the regulative cultural narratives and norms of a time. By saying "disembodied human fantasy" it is alluded to a strong degree of subjectivism that is in a relatively low degree of attunement with the needs of the environment. In other words, disembodiment means that the innovative idea of a subject is not embedded in a lived and subjective body empathizing with the needs of the environment. Instead, it springs from the innovator's personal fantasy. The source of that fantasy, if it is not rooted in empathy, seems to be the cultural narrative of a time and/or industry. In our time and in the tech industry this is often science-fiction (Laakasuo et al., 2018).

75 The Kantian term of "regulative idea" is used here, because people act "as if" the technological ideas and innovations were set to determine the future, and thereby influence or even regulate behavior, such as investment behavior, education, work priorities, etc.

76 A 2019 study by Prof. Nik Franke and his colleagues at WU Vienna showed that decision-makers across nine industries underestimated the share of bottom-up user innovations to range only from 16–34% of all innovations in their industry where in truth they made up 44–87% of the products and services we consume today (Bradonjic et al., 2019).

77 Strategic considerations, which do not spring from a real natural need, run the risk of fostering technology determinism; a determinism that then attains no more than "acceptance" from future customers. It might not be surprising that technology scholars rarely speak of users embracing or cherishing a technology, but always use the terminology "Technology Acceptance." From a linguistic perspective this is quite revealing. "Accepting" something is very different from needing or wanting it, as the Technology Acceptance Literature has amply shown (Venkatesh, Morris, Davis, & Davis, 2003). One could argue that acceptance is more comparable with "swallowing" something than actually enjoying it. And here precisely resides the difference between pure-will innovators and real-value innovators. The pure-will innovators "design the world as it pleases them," involving "users" to optimize acceptance. Real-value innovators, in contrast, truly need something for themselves and build it for themselves. That said, the Design Thinking Schools explicitly start their projects (at least in theory) with a phase they call "emphasize." This emphasizing seeks very concretely to ensure that technology or innovations inserted into a market are based on needs and not just swallowed.

78 An interesting question is whether eros can be misled. In this book pure-will innovations is reflected on and it is easy to confound eros with pure will. Is eros not something that drives our pure will? Eros should be a striving for positive value potentials happening in tune with the world, while pure will is a largely disembodied fantasy. In pure-will innovation the inventor is fantasizing in his own mind about what should be desired without "being in the world".

79 Note that this use of the word "milieu" in terms of individual resonance space is slightly different from a social "milieu" that may exist in community like Silicon Valley.

Abbreviations

COT	Commercial Of-the-shelf Technology
CSR	Corporate Social Responsibility
CV	Curriculum Vitae
IS	Information Systems
EVR	Ethical Value Requirement
GDP	Gross Domestic Product
SDLC	System Development Life Cycle
SLA	Service Level Agreement
SOI	System of Interest
SOS	System of Systems
VBE	Value-Based Engineering

Index

Abraham Maslow 87
Affordance 42
AI V, VII, 7, 9–10, 11, 19–21, 26–27, 69, 72–73, 81, 83, 100, 113–114, 122, 125–126, 136, 138–139, 156, 174, 176–177
Alasdair MacIntyre 92
Albertus Magnus 1
algorithm design 20, 72
algorithms 6, 27, 66
Aristotle 87
artificial intelligence 6
attention 5, 13, 49–50, 171
attentiveness 24
automation 1–2, 77
axiology 58–59

beauty 42, 52–55, 58, 62, 157
Benjamin Franklin 89, 110
best-available-technique 119
Big Data 11, 72, 139, 177
Buddhism 87
bushidō code 88

care VII, 8, 11, 13, 18, 21, 23–24, 36, 61, 84–85, 104, 112, 116, 126, 138, 171
Clayton Christensen 146
Clyde Kluckhohn 42–43
commercial off-the-shelf services 71
Concept of Operations 9, 23, 60, 65, 67, 75, 81, 87, 97, 99, 121, 126, 132–133, 156
Confucian ethics 87
ConOps 9, 174
context 3–4, 8–9, 11, 16, 20, 25–27, 34, 36, 47–48, 55, 57–59, 61, 63–66, 68, 70, 77–78, 80–83, 91, 100, 109, 126–127, 133–134, 136, 138, 152–153, 169, 173–174
control V, VI, 5–6, 11, 18–20, 27–28, 34, 36, 38–39, 63, 71–72, 80, 100, 109, 118–122, 126, 130–132, 135, 174, 176
controller 69
core values 31, 37, 52, 57, 100–102, 106, 109, 157, 159
CSR 33, 36, 85–86, 107

Data flows 9, 16, 27, 34, 60, 67–68, 83, 133, 159, 174
democracy V, 5
Design Thinking 65, 143–144, 173, 177
dignity V, 3, 13, 16–17, 58, 62, 94–95, 126, 169, 175
Direct stakeholder 65
discourse ethics 77
Disruptive Innovation 146
dodecahedron 9, 160
Don Ihde 90
Duty Ethics 24, 93

ecosystem responsibility 18–20
Edmund Husserl 52, 170
efficiency VII, 8, 10, 66, 70, 106, 108, 125, 148
Eros 144
ethical feasibility 61, 63, 79–80, 173
Ethical Policy Statement 32–33, 110
Ethical System Design 17
Ethical Value Requirements VIII, 4, 9, 37, 86, 111–112, 175
Ethically Aligned Design 39, 111, 114
Ethics 82, 87, 100, 102, 133
Eudaimonia 87
Evagrius Ponticus 99, 169, 175
Evil 59
EVR VIII, 38, 111–120, 129–130, 135, 176
EVRs 4, 37–39, 86, 111–116, 118–120, 129, 132, 135, 156, 175
extrinsic 58
extrinsic values 58

feasibility analysis 60, 79–80
flourishing 3
foundation model 72
Francis Bacon 1
freedom V, 3, 10, 13, 22, 52, 92, 125–126, 169

garden 3, 34, 39, 60, 63, 68, 103, 160, 173
GDP 1–2, 5, 12
GDPR V, 9, 110
General Data Protection Regulation V, 9, 110
Gestalt 41, 50, 172

Giotto 87–88
Good 59
growth 2, 12, 30, 96, 156

Hannah Arendt 94–95
health V, VII, 1, 3, 7, 10, 16–17, 38, 66, 68–69, 77, 98, 100, 105, 109, 112–114, 122, 125, 127, 130
humanity 1, 4, 6, 10–11, 90, 95, 136, 138, 141, 148, 171
humbleness 6, 24, 52, 62

ideal speech situations 78
IEEE 7000™ 8, 10, 14, 17, 19–21, 22, 23, 24, 25, 31, 33–34, 39, 57, 64, 71–72, 78, 80–83, 85–87, 90–91, 96, 100–102, 105, 108, 111–115, 118–119, 121–122, 126, 132–133, 170, 174–176
Ikujiro Nonaka 31, 63, 152
impact analysis 127
impact assessment 39, 100, 114, 123, 125–126, 131, 172, 176
impact-assessment 123
indirect stakeholder 22, 77, 84
innovation VI, VII, 1, 3–4, 7–8, 10–13, 22, 26, 30, 32, 36, 39, 63, 65, 78, 81, 84, 86, 96, 136–137, 141–147, 149–150, 152–153, 173, 177
Intrinsic values 57
ISO/IEC/IEEE 15288 19, 175
ISO/IEC/IEEE 24748-7000 VI, 8

Jeremy Bentham 91
John Stuart Mill 91
junzi 87

Kant 93–94, 106
kindness 24, 87
knowledge VII, 3, 6, 31, 36, 40–41, 47–50, 52, 56–57, 62–63, 77, 83–84, 86–87, 100, 105, 130, 136, 147–148, 151–153, 173–174

Laozi 13
lead user 142, 144, 151–153, 173
life-form 47

Material Value Ethics VII, 40, 107–108, 171, 175
Max Scheler VII, 40–41, 46, 58–59, 152, 171
Michael Porter 30

milieu 11, 14, 42, 47–48, 51, 63–64, 150, 152–153, 157, 173, 177
Minimum Viable Product 121
moderation 24, 87–89
moral consciousness 41
moral philosophy 25, 82, 91, 93
Morality 82

needs 9, 11, 36, 39, 51–53, 61–62, 65, 71, 74, 77–78, 82, 99, 115, 118, 120, 122, 127, 138, 144, 149–150, 173–174, 177
NGO 28, 81, 84
Nicolai Hartmann 41–42, 52, 170
NIST standard for system security 39

observability 71
ontology VII, 52–53, 58–59, 170, 172

Performative Acts 152
personal data 5, 17, 36, 70, 80, 100, 104–107, 111, 113, 117, 119, 129
personas 76
phronimoi 87
Platon 157
Prioritizing 102
privacy V, 10, 13–17, 18, 19, 23, 26–27, 36, 38, 40–41, 43–44, 52, 56–59, 61–62, 70, 73, 80, 83, 93, 100–101, 104–105, 107, 109, 111, 113, 118–123, 126–127, 129, 132, 174, 176
productivity VII, 2, 10, 30, 93, 96
progress 1–2, 3, 4, 12, 89, 136, 138, 140, 142, 150
protection demand 123, 127, 129, 131, 135
Pure-Will Innovation 138
purpose VIII, 3, 14, 30, 58, 62, 103, 120, 144, 155, 176

Real-Value Innovations 142
regress 1–4, 7
reliability 6, 17, 24, 38, 42, 45, 74, 105, 120, 148
requirements' engineering 13
Residual risks 132
responsibility 14, 18–19, 27, 29, 36, 60, 70, 72, 84–85, 101, 105, 107, 133, 171
RFID 61–62, 73, 127, 132, 138
Risk Management 118
risk-based 8, 38–39, 114, 116–120, 122–123, 126, 131–132, 135, 170, 176
Risk-based Design 114, 122

science-fiction 100, 177
security V, 3, 14–15, 16, 17, 23, 27, 36, 38, 47–52, 58, 70, 74–75, 83, 93, 100, 109, 114, 118–119, 122–123, 126–127, 129, 142, 157, 169, 172
service agreements 6
service level agreement 71
SLA 71
social dilemma 5
software VI, VII, 6–7, 28, 47, 65, 114, 117–118, 122, 124, 129, 152, 158–160, 169, 176–177
SOI 8–9, 10, 17, 25, 34, 36, 39, 46, 60–61, 63–84, 86–87, 90, 95, 97–98, 100, 102, 105–107, 110–111, 114–115, 118–121, 131–135, 160, 173, 175–176
SOI views 74–75, 81, 83
SOS 17–19, 68–74, 76, 78, 81, 83, 87, 133, 173
stakeholder dialogue 21, 102
Stakeholder representatives 76, 84
surveillance capitalism 5
system design V, VI, 7–8, 9, 10, 16, 18, 39, 44, 58–59, 65–66, 76, 81–82, 85–86, 102, 110, 114, 118, 122–123, 125–126, 130–131, 135, 176
system engineering 9, 13, 52, 84, 144, 170
System of Interest 8, 17
System of Systems 17

tacit knowledge 49
technological determinism 137
tetrahedron 37, 52, 157
threats 86, 116–119, 123, 129–132, 135
Timaios 157–158
training data 20
Transparency Management 8, 78

Umberto Eco 55
UML 34, 67, 169, 174
Utilitarianism VI, 24–25, 87, 90–93, 96, 110

value V, VII, 1–2, 3, 7–13, 15, 17–19, 23–34, 36–59, 61–66, 69–70, 73, 75–77, 79–87, 91, 93–123, 126–136, 138, 140, 142, 144–150, 152–153, 157–160, 169–177
value bearer 53, 145, 171

value clusters VIII, 26, 36, 57–59, 86, 101–103, 106, 110
value demonstrator 57, 101
value demonstrators 57, 119, 170
value disposition 47–48, 119, 153, 158, 170
value exploration 25–26, 36, 65, 76, 81–85, 127
Value Lead 85–87, 97–99, 101–102, 110, 120, 127, 134
Value principle lists 27
value qualities VII, 3, 10, 26, 36–38, 45, 49–52, 54–59, 64, 83, 100–103, 106, 109, 111–113, 115, 118–119, 122, 126–127, 129, 131–132, 134, 145, 148, 152–153, 157–159, 170–173
Value Register 33–34, 102, 112–113, 121, 132–135
Value Sensitive Design V, 25, 65, 170
Value-Based Engineering V, VIII, 8–9, 15–17, 19–21, 23, 25, 28–29, 32, 34–35, 37, 44, 52, 56, 60, 66, 78, 81–82, 87, 118, 126, 136, 142, 150, 153, 156, 158, 170, 173
valueception 3, 46
values V, VII, 37–8, 9, 10, 11, 13–14, 16–19, 21, 23–26, 29, 31–34, 36–38, 40–47, 50, 52–55, 57–59, 61–62, 66, 78, 81–86, 89, 92–93, 95–96, 98–102, 105–111, 114, 116, 118–119, 122, 125–126, 129–132, 134–135, 142–143, 146, 148–150, 153, 156–158, 169–173, 175–176
VBE 8–9, 10, 11, 13, 16, 20–21, 23, 26, 28, 35–36, 39, 60–65, 67, 70–71, 74–75, 78–85, 90–91, 93, 95–97, 100–102, 105–107, 109–111, 114–116, 118121–122, 126–127, 129–133, 136–137, 143, 145, 153, 159–160, 171–173, 175–176
Virtue Ethics 24, 87
virtues 24, 87–90, 95–96, 100, 136

wellbeing 1–2, 3, 8, 11, 26–27, 106, 122, 148, 169
Werner Ulrich 84
William David Ross 95
Wittgenstein 47, 170
Wolfie Christl 18

Zuboff 5, 13, 113

www.ingramcontent.com/pod-product-compliance
Lightning Source LLC
Chambersburg PA
CBHW080539300426
44111CB00017B/2801